Lyman Spalding Foster

A Consideration of Some Ornithological Literature

With Extracts from Current Criticism - I. 1876-1883 II: 1884 to 1893

Lyman Spalding Foster

A Consideration of Some Ornithological Literature
With Extracts from Current Criticism - I. 1876-1883 II: 1884 to 1893

ISBN/EAN: 9783337241544

Printed in Europe, USA, Canada, Australia, Japan

Cover: Foto ©berggeist007 / pixelio.de

More available books at **www.hansebooks.com**

AUTHOR'S EDITION.

EXTRACTED FROM

Abstract of the Proceedings of the Linnæan Society of New York.

No. 6, 1894, pp. 47–99.

A Consideration of Some Ornithological Literature, with Extracts from Current Criticism.

I. 1876 TO 1883.

II. 1884 TO 1893.

By L. S. Foster.

1876 to 1883.

Under the heading, "Recent Literature," in the volumes of the Bulletin of the Nuttall Ornithological Club, published from 1876 to 1883, are reviews of numerous publications which, I hold, pretty fairly represent the ornithological literature of this important period, particularly so far as North America is concerned.

A hasty survey of this literature might, perchance, convey the idea of individual effort rather than combined exertion, but, summarized, it shows an advanced movement along a series of lines which, at the close of the period, interlaced and formed the firm foundation upon which has been erected the solid superstructure of The American Ornithologists' Union.

The more prominent features of the time and those which will permanently characterize it, seem to be as follows:

The appearance of the first volume of the Catalogue of the Birds in the British Museum;

The revisionary work of Mr. Ridgway and Dr. Stejneger on certain orders and genera;

The organization of The Linnæan Society of New York, together with the publication of the first volume of its Transactions;

The publication of Biologia Centrali-Americana;

The appearance of Mr. George N. Lawrence's "General Catalogue of the Birds noted from the Islands of the Lesser Antilles"; Stearns and Coues's "New England Bird Life"; Dr. Merriam's "Review of the Birds of Connecticut"; Dr. Wheaton's "Report on the Birds of Ohio"; Dr. Coues's series of four bibliographical papers and his check-list of 1882; Mr. Ridgway's nomenclature of 1881; and the beginning of John Burroughs's charming series of out-of-door books with the republication of "Wake-Robin" in 1877.

In clearing the way for the A. O. U. check-list, the work on nomenclature which was done by Mr. Ridgway and Dr. Stejneger was not only necessary but eminently workmanlike. In these years, the battle for trinomialism in North America was fought and gallantly won.

Especially will this period be notable as the epoch in which serious work was begun in recording facts of migration: the Germans, the English, and, in this country, Prof. W. W. Cooke, accomplished much.

[The continuation of this paper, as read, consisted of numerous examples of the criticisms which follow:]

TITLES AND CRITICISMS OF

SOME ORNITHOLOGICAL LITERATURE,

I.

1876 TO 1883.

1876.

COOPER, J. G.—New Facts relating to Californian Ornithology. No. 1. By Dr. J. G Cooper. *Proc. Cal. Acad. Sci.*, 1876. 14 pages.

....About fifty species are noticed....The paper is replete with interesting matter, and forms a valuable contribution to our knowledge of Californian Ornithology.—J. A. A., *Bull. Natt. Ornith. Club*, Vol. II., p 76, July, 1877.

D'Hamonville, J. C. L. T.—Catalogue des Oiseaux d'Europe, ou énumération des espèces et races d'oiseaux dont la présence, soit habituelle soit fortuite, a été dûment constatée dans les limites géographiques de l'Europe, par J. C. L. T. D'Hamonville. 8vo., pp. 74. Paris, 1876.

....deserves more than a mere mention on account of the admirably comprehensive manner in which it has been prepared. the Baron makes the whole number 658,....—T. M. B., *Bull. Nutt. Ornith. Club*, Vol. II., pp. 106, 107, October, 1877.

Garrod, A. H.—On some Anatomical Characters which bear upon the Major Divisions of the Passerine Birds. By A. H. Garrod. *Proc. Zoöl. Soc. London*, 1876.

....He concludes his paper with a tabular arrangement of the larger groups of the Passeres, expressive of his views of their affinities. —J. A. A., *Bull. Nutt. Ornith. Club*, Vol. II., p. 23, January, 1877.

Gentry, Thomas G.—Life-Histories of the Birds of Eastern Pennsylvania. By Thomas G. Gentry. (In two volumes.) Vol. I: Philadelphia, 1876. 12 mo., pp. xvi., 309.

... a most welcome volume of biographies of the birds of Eastern North America ....The author's style is unostentatious and simple, at times lapsing into carelessness ...The present volume includes the Song-birds as far as the Corvidæ of Dr. Coues's arrangement ...—J. A. A., *Bull. Nutt. Ornith. Club*, Vol. I., pp. 49, 50, July, 1876.

Henshaw, H. W.—Annual Report upon the Geographical Surveys West of the One-Hundredth Meridian, etc. By George M. Wheeler, First Lieutenant of Engineers, U. S. A. Being Appendix J J of the Annual Report of the Chief of Engineers for 1876. Washington: Government Printing Office, 1876. Report on the Ornithology of the Portions of California visited during the Field Season of 1875. By Mr. H. W. Henshaw. Pp. 224-278.

....Among the more important results are the extension, either southward or westward, of the previously recorded range of many species of birds ...The biographical annotations are often full and always exceedingly interesting ...—W. B., *Bull. Nutt. Ornith. Club*, Vol. III., pp. 136, 137, July, 1878.

Jordan, David Starr.—Manual of the Vertebrates of the Northern United States, including the District east of the Mississippi River, and north of North Carolina and Tennessee, exclusive of Marine Species. By David Starr Jordan, M.S., M.D., etc. Chicago, 1876. 12mo., pp. 342.

....Several of the analytical tables of different groups of birds are based on or taken directly from Coues's Key, and the latest and best authorities are followed for the other classes .. On the whole the author is to be congratulated on the success he has achieved in this difficult undertaking, combining in a work of convenient size and moderate cost a text-book of the Vertebrate Animals of the Northeastern States reliable in character and sufficiently extended to guide the student with tolerable ease to the name of any species he may chance to have in hand. J. A. A., *Bull. Nutt. Ornith. Club*, Vol. I., pp. 93, 94, November, 1876.

KIDDER, J. H.—Contributions to the Natural History of Kerguelen Island. By J. H. Kidder, M.D. Edited by Dr. Elliott Coues, U. S. A. II. Oölogy, pp. 6–20. *Bull. U. S. Nat. Mus.*, No. 3. Washington, 1876.

....an account of the Oölogy of the island, including detailed descriptions and measurements of the eggs, together with an account of the breeding habits of all the species found breeding there.... J. A. A., *Bull. Nutt. Ornith. Club*, Vol. I., p. 48, July, 1876.

KIDDER, J. H. Contributions to the Natural History of Kerguelen Island. By J. H. Kidder, M.D. Edited by Dr. Elliott Coues, U. S. A. II., pp. 85–116. A Study of *Chionis minor*. *Bull. U. S. Nat. Mus.*, No. 3. Washington, 1876.

This essay opens with a résumé of the literature of the species. Then follows a description of its anatomy, including an account of its myology, of the viscera and the skeleton; of its habits, general appearance in life, and external characters.... J. A. A., *Bull. Nutt. Ornith. Club*, Vol. I., pp. 48, 49, July, 1876.

LAWRENCE, GEORGE N.—Description of a New Species of Jay of the Genus Cyanocitta; also of a supposed New Species of Cyanocorax. By George N. Lawrence. *Annals of the Lyceum of Nat. Hist. N. Y.*, Vol. XI., pp. 163–165. [Published Feb., 1876.]

....(*Cyanocitta pulchra*) being from Ecuador and the other (*Cyanocorax orloni*) from Northern Peru. J. A. A., *Bull. Nutt. Ornith. Club*, Vol. I., p. 47, July, 1876.

LAWRENCE, GEORGE N.—Birds of Southwestern Mexico collected by Francis E. Sumichrast. Prepared by George N. Lawrence. *Bull. U. S. Nat. Mus.*, No. 4. Washington, 1876.

. The list embraces three hundred and twenty-one species, with valuable and occasionally quite copious field-notes by the collector .. J. A. A., *Bull. Nutt. Ornith. Club*, Vol. I., p. 93, November, 1876.

MARSH, O. C.—Extinct Birds with Teeth. By Professor O. C. Marsh. *Am. Jour. Sci. and Arts*, June, 1876, pp. 509–511.

These interesting forms ... combine in a peculiar manner many reptilian characters with others truly avian. J. A. A., *Bull. Nutt. Ornith. Club*, Vol. I., p. 49, July, 1876.

RIDGWAY, ROBERT.—Second Thoughts on the Genus Micrastur. By Robert Ridgway. *The Ibis*, 1876, pp. 1–5.

RIDGWAY, ROBERT.—Studies of the American Falconidæ: Monograph of the Polybori. By Robert Ridgway. *Bull. U. S. Geol. and Geogr. Surv. of Terr.*, Vol. I., No. 6, pp. 451-473, plates xxii.-xxvii., February 8, 1876.

RIDGWAY, ROBERT.—Studies of the American Falconidæ. By Robert Ridgway. *Bull. U. S. Geol. and Geogr. Surv. of Terr.*, Vol. II., No. 2, pp. 91-182, plates xxxi., xxxii., April 1, 1876.

SAUNDERS, HOWARD.—On the Stercorariinæ or Skua Gulls. By Howard Saunders, F.L.S. &c. *Proc. Zoöl. Soc. London*, 1876, pp.317-332, pl. xxiv.

Mr. Saunders recognizes six species, all of which he refers to one genus for which he adopts the name *Stercorarius*....—J. A. A., *Bull. Nutt. Ornith. Club*, Vol. II., pp. 23, 24, January, 1877.

SAUNDERS, HOWARD.—On the Sterninæ, or Terns, with Descriptions of three new Species. By Howard Saunders, F.L.S., F.Z.S. *Proc. Zoöl. Soc. London*, 1876, pp. 638-672, pl. lxi.

... Of the forty-eight species recognized, thirty-eight are placed under *Sterna*....—J. A. A., *Bull. Nutt. Ornith. Club*, Vol. II., p. 24, January, 1877.

....We have here in condensed and convenient shape the main results of a protracted study, representing much laborious and faithful application; the author has evidently worked with care, and fully availed himself of the unusual facilities he has enjoyed....I regard the paper as the most authoritative one we possess on this subject. ....The colored plate illustrates the heads of three species of *Anous* ... —Elliott Coues, *Bull Nutt. Ornith. Club*, Vol. III., pp. 140-144, July, 1878.

SCLATER, P. L. and SALVIN, OSBERT.—On new Species of Bolivian Birds. By P. L. Sclater, M.A., Ph.D., F.R.S., and Osbert Salvin, M.A., F.R.S. *Proc. Zoöl. Soc. London*, 1876, pp. 352-358, pll. xxx-xxxiii.

SCLATER, P. L. and SALVIN, OSBERT.—Revision of the Neotropical Anatidæ. By P. L. Sclater and O. Salvin. *Proc. Zoöl. Soc. London*, 1876, pp. 358-412, pl. xxxiv.

...a most valuable synopsis of the Ducks and Geese of Middle and Southern America, and embraces also a large proportion of the species of North America, including as it does all that reach Tropical America in their migrations .. The paper closes with a very convenient tabular synopsis of the geographical distribution of the genera and species. —J. A. A., *Bull. Nutt. Ornith. Club*, Vol. II., p. 24, January, 1877.

VENNOR, HENRY G.—Our Birds of Prey; or the Eagles, Hawks, and Owls of Canada. By Henry G. Vennor, F.G.S. Of the Geological Survey of Canada. With 30 Photographic Illustrations by Wm. Notman. Montreal: Published by Dawson Brothers. 1876. 4to., pp. i-viii and 1-154, 30 mounted photographs of birds.

....The text, which is largely compiled from the notes of other writers, gives a fairly digested summary of the individual history of each species .. T. M. B., *Bull. Nutt. Ornith. Club*, Vol II., pp. 24 25, January, 1877.

1877.

BARROWS, W. B.—Catalogue of the Alcidæ contained in Museum of the Boston Society of Natural History, with a review and proposed classification of the Family. By W. B. Barrows. *Proc. Boston Soc. Nat. Hist.*, Vol. XIX., pp. 150 165. November, 1877.

...The true affinities of the species he (Mr. Barrows) believes can only be determined by a thorough study of their embryological development. The character of this paper indicates that in Mr. Barrows we have a valuable accession to our corps of ornithological students. J. A. A., *Bull. Nutt. Ornith. Club*, Vol. III., p. 86, April, 1878.

BENDIRE, CHARLES E.—Notes on some of the Birds found in Southeastern Oregon, particularly in the Vicinity of Camp Harney, from November, 1874, to January, 1877. By Captain Charles Bendire, U. S. Army. *Proc. Boston Soc. Nat. Hist.*, Vol. XIX., pp. 109-149, Nov., 1877.

... a list embracing one hundred and ninety-one species and varieties .. Aside from some former notes by the same author....we have here our first detailed information respecting the ornithology of the immediate region under consideration .. The list is enriched with copious biographical notes, including descriptions of the breeding-habits, nests, and eggs of a large number of the less well-known species, and forms a most important contribution to the ornithology of the West. J. A. A., *Bull. Nutt. Ornith Club*, Vol. III., p. 81, April, 1878.

BURROUGHS, JOHN.—Wake-Robin. By John Burroughs. Second Edition, corrected, enlarged and illustrated (cut). New York: Published by Hurd and Houghton. Cambridge: The Riverside Press, 1877, 16mo., pp. 1 256, frontispiece and wood cuts.

Hurd and Houghton have reprinted Mr. John Burroughs's charming little volume "Wake-Robin," wherein the wild wood-life of the birds, from Washington to the Adirondacks is picturesquely sketched. Mr. Burroughs has a keen eye and a loving heart towards the birds .... -E. I., *Bull. Nutt. Ornith. Club*, Vol. II., pp. 18, 19. April, 1877.

ELLIOT, D. G.—Review of the Ibidinæ, or Subfamily of the Ibises. By D. G. Elliot, F.R.S.E., F.L.S., etc., etc. *Proc. Zool. Soc. London*, 1877, pp. 477 510, pl. li.

....Mr. Elliot treats the Ibises and Spoonbills as subfamilies of one family, for which he adopts the name *Ibididæ*. After a short *résumé* of the literature of the subject he gives a key to the nineteen genera (three being new), among which he distributes his twenty-five species. Then follows a systematic review of the species, with their principal synonomy, and various critical and descriptive remarks with generally a short account of their habits and geographical distribution.... J. A. A., *Bull. Nutt. Ornith. Club*, Vol. III., p. 182, October, 1878.

FEILDEN, H. W.—List of Birds observed in Smith Sound, and in the Polar Basin during the Arctic Expedition of 1875–76. By H. W. Feilden. *The Ibis*, Fourth Series, Vol. I., pp. 401–412, October, 1877.

....enumerates twenty-four species observed.... "in Smith Sound and northward between the seventy-eighth and eighty-third degrees of north latitude,"....The quite detailed notes respecting the species of this list render it a paper of unusual interest. —J. A. A., *Bull. Nutt. Ornith. Club*, Vol. III., p. 86, April, 1878.

GENTRY, THOMAS G.—Life-Histories of the Birds of Eastern Pennsylvania. By Thomas G. Gentry. Vol. II., 8vo., pp. 336. The Naturalist's Agency, Salem, Mass., 1877.

....It abounds in original observations, combined with much that is gleaned from other authors....Despite some faults of execution, the work before us contributes much of value respecting the habits of our birds, and records many interesting points in their history not given by previous writers.—J. A. A., *Bull. Nutt. Ornith. Club*, Vol. III., pp. 36, 37, January, 1878.

HARVIE BROWN, J. A.—On the Distribution of Birds in North Russia. Part I. On the Distribution of Birds of the Lower Petchora, in Northeast Russia. Part II. Longitudinal Distribution of Species North of 64° 30′ N. lat., or the Northern Division. Part III. On the Longitudinal Distribution of the Birds of the Southern Division (between 64½° N. and 58°–60° N.). By J. A. Harvie-Brown. *Annals and Magazine of Natural History*, April, July, and September, 1877.

....By means of a system of symbols the range of each of the two hundred and eighty-one positively identified or authentic species is given in tables, in such a way as to indicate the abundance or scarcity of the species in each of the several districts ..It is good work in a most important direction ...The number of circumpolar species (nearly fifty) embraced in these lists render these papers of special interest to students who commonly confine their attention to the birds of the North American Region.—J. A. A., *Bull. Nutt. Ornith. Club*, Vol. III., pp. 35, 36, January, 1878.

HENSHAW, H. W.—Annual Report upon the Geographical Surveys West of the One-Hundredth Meridian, etc. By George M. Wheeler, First Lieutenant of Engineers, U. S. A. Being Appendix N N of the Annual Report of Engineers for 1877. Washington: Government Printing-Office, 1877. Report on the Ornithology of Portions of Nevada and California. By Mr. H. W. Henshaw. Pp. 1303–1322.

....following is a systematic and very able consideration of the faunal provinces of the United States....The full results of the season's work are given in two detailed lists, entitled, respectively, "List of Birds observed near Carson City, Nevada, from August 25 to September 16, and from November 10 to November 20, 1876, with Notes," and "List of Birds observed on the Eastern Slope of the Sierras, near Carson City, Nevada, from September 16 to November 7, with Notes." .. The genus *Passerella* is again overhauled.... -W. B., *Bull. Nutt. Ornith. Club*, Vol. III, pp. 137, 138, July, 1878.

LANGDON, FRANK W.—A Catalogue of the Birds of the vicinity of Cincinnati, with Notes. By Frank W. Langdon. Salem, Mass. The Naturalist's Agency, 1877, 8vo., pp. 18.

... embraces two hundred and seventy-nine species, about one third of which are marked as known to breed in the vicinity... The list is evidently prepared with care, and gives a convenient and undoubtedly trustworthy summary of the Avian Fauna of the locality of which it treats.- J. A. A., *Bull. Nutt. Ornith. Club*, Vol. III., p. 34, January, 1878.

LAWRENCE, GEORGE N.—Descriptions of New Species of Birds from the Island of Dominica. By George N. Lawrence. *Ann. N. Y. Acad. Sci.*, Vol. I., pp. 46-49. Issued Dec., 1877.

The important explorations by Mr. F. A. Ober in some of the smaller West India Islands (Lesser Antilles have been rich in interesting results relating to birds. The collections and observations made by Mr. Ober have been made the basis of several recent papers by Mr. George N Lawrence, in which no less than fourteen species supposed to be new have been described . J. A. A., *Bull. Nutt. Ornith. Club*, Vol. IV., pp. 48, 49, January, 1879.

McCAULEY, C. A. H.—Notes on the Ornithology of the Region about the Source of the Red River of Texas, from Observations made during the Exploration conducted by Lieutenant E. H. Ruffner, Corps of Engineers, U. S. A. By C. A. H. McCauley, Lieutenant Third United States Artillery. Annotated by Dr. Elliott Coues, U. S. A. *Bull. U. S. Geol. and Geogr. Surv. of Terr.*, Vol. III., No. 3, pp. 655-695, May 15, 1877.

... The paper includes notices of about one hundred species, with quite copious notes respecting the habits of a considerable proportion of them, with, in some cases, descriptions of their nests and eggs.... --J. A. A, *Bull. Nutt. Ornith. Club*, Vol. II., pp. 76, 77. July, 1877.

MERRIAM, C. HART.—A review of the Birds of Connecticut, with Remarks on their Habits. By C. Hart Merriam. *Trans. Conn. Acad. of Arts and Sciences*, Vol. IV., pp. 1-150, July-Oct., 1877.

....Since the appearance of Linsley's "Catalogue of the Birds of Connecticut" in 1843, no detailed enumeration of the birds of that State has been published. Hence the advent of Mr. Merriam's paper must be hailed with interest by all engaged in the study of New England Ornithology. The author gives in all two hundred and ninety-two species . In the careful elaboration of interesting details culled from personal experience and the note-books of well-known and trustworthy field collectors, this paper is most rich .. W. B., *Bull. Nutt. Ornith. Club*, Vol. II., pp. 107, 108, October, 1877.

MINOT, H. D.—The Land-birds and Game-birds of New England, with descriptions of the birds, their nests and eggs, their habits and notes. With illustrations. By H. D. Minot.

"To him who in the love of Nature holds
"Communion with her visible forms, she speaks
"A various language;"
Bryant's Thanatopsis.

Salem, Mass. Naturalists' Agency. Boston: Estes & Lauriat. 1877. 8vo., pp. i-xvi and 1-456, frontispiece and woodcuts.

....the descriptions, however, are tersely original... the most prominent and most original features of the work are the artificial "keys."....—E. C., *Bull. Nutt. Ornith. Club*, Vol. II., pp. 49, 50, April, 1877.

.. the book has never been reviewed on its merits, and things which should have been severely censured have passed nearly unchallenged up to the present time .. leaving out the faulty portions, which in nearly all cases relate to abstract points similar to those just cited [careless methods of work and identification], the pages bear the impress of accurate observation and original thought, while no one who loves the out-door side of Nature can fail to sympathize with the author's sentiment or to be impressed by the truth and beauty of many of his passages It is a pity that one who writes so delightfully will mar his work by a persistent adhesion to false principles.—William Brewster, *Bull. Nutt. Ornith. Club*, Vol. VI., pp. 242-244, October, 1881.

NELSON, E. W.—Birds of Northeastern Illinois. By E. W. Nelson. *Bull. Essex Inst.*, Vol. VIII., pp. 90-155, April, 1877.

It is not, however, from the simple enumeration of species that this list derives its chief value and interest, but from the unusually complete and satisfactory character of the biographical annotations, which embrace good descriptions of the habits of many birds previously but little known ...—W. B., *Bull. Nutt. Ornith. Club*, Vol. II., pp. 68, 69, July, 1877.

NELSON, E. W.—Notes upon Birds observed in Southern Illinois, between July 17 and September 4, 1875. By E. W. Nelson. *Bull. Essex Inst.*, Vol. IX., pp. 32-65, June, 1877.

...contains much information respecting the distribution, habits, and relative abundance of the summer birds of the southern portion of the ...State....—J. A. A., *Bull. Nutt. Ornith. Club*, Vol. III., p. 36, January, 1878.

RATHBUN, FRANK R.—A Partial Catalogue of the Birds of Central New York, from observations taken in the Counties of Cayuga, Seneca, and Wayne by Mr. H. G. Fowler, of Auburn, N. Y., and from the Cabinet of Skins of New York Birds collected by Mr. J. B. Gilbert, of Penn Yan, Yates County. Divided and arranged in accordance with the "Check List of North American Birds," by Elliott Coues, M.D., U. S. A., and dedicated to the Cayuga Historical Society. By Frank R. Rathbun. *Auburn Daily Advertiser* (newspaper) of August 14, 1877.

....The list contains one hundred and ninety-one species, with brief notes on their relative abundance, times of migration, etc. The list bears evidence of trustworthiness . — J. A. A., *Bull. Nutt. Ornith. Club*, Vol. III., pp. 34, 35, January, 1878.

Reichenow, Anton.—Systematische Uebersicht der Schreitvögel (Gressores), einer natürlichen, die Ibidae, Ciconiidae, Phoenicopteridae, Scopidae, Balaenicipidae, und Ardeidae umfassenden Ordnung. Von Dr. Ant. Reichenow, Assistent am kgl. zoolog. Museum in Berlin. *Journal für Ornithologie*, XXV Jahrgang, pp. 113 171, 225 278, pll. i, ii. April and July, 1877.

....He also throws over all "barbarous" names, whether specific or generic, all names of erroneous signification, and all classical names improperly constructed. Under these restrictions many long-established and familiar designations fall, to be replaced by the next (in Dr. Reichenow's view) unobjectionable name. In default of any such our author proceeds to supply the deficiency. While differing from Dr. Reichenow respecting important principles of nomenclature, and on various points of classification, we can but accord to his paper a high importance, as it evinces laborious and careful research, and embraces a vast amount of information, succinctly and lucidly presented, that will be of great service to future workers in the same field. J. A. A., *Bull. Nutt. Ornith. Club*, Vol. III., pp. 183 185, October, 1878.

Ridgway, Robert.—Report of Geological Explorations of the Fortieth Parallel. Clarence King, Geologist in Charge. Vol. IV., Part III., Ornithology. By Robert Ridgway. 4to., pp. 303 670. 1877.

....a thorough and exhaustive account of the ornithology of an interesting belt of country. The observations were mainly limited to that portion of the Great Basin included between the thirty-ninth and forty-second parallels and extending from the Sierra Nevada to the Wahsatch Mountains . in point of nomenclature it represents the author's later views. J. A. A., *Bull. Nutt. Ornith. Club*, Vol. III., pp. 82, 83 April, 1878.

Roosevelt, Theodore, Jr., and Minot, H. D.—The Summer Birds of the Adirondacks in Franklin County, N. Y. By Theodore Roosevelt, Jr., and H. D. Minot. 8vo., pp. 4, 1877.

....a very acceptable list of the summer birds of the Adirondacks embracing ninety-seven species. J. A. A., *Bull. Nutt. Ornith. Club*, Vol. III., p. 36, January, 1878.

By far the best of these recent (local) lists which I have seen....—C. H. M., *Bull. Nutt. Ornith. Club*, Vol. III., p. 85, April, 1878.

Rowley, G. D.—Somateria labradoria (J. F. Gmelin). The Pied Duck. By G. D. Rowley, M.A., F.L.S., F.Z.S., etc., etc. Ornithological Miscellany, Vol. II., Part VI., pp. 205 223, with 5 plates, 1877. London: B. Quaritch, 15 Piccadilly, W.; Trübner & Co., Ludgate Hill, E. C.; R. H. Porter, 6 Tenterden St., Hanover Square, W.

... a timely and exhaustive contribution to the history of a species believed to be rapidly approaching extinction.... Mr. Rowley here gives not only the literary history of the species, but discusses its relationship to the Eiders ...—J. A. A., *Bull. Nutt. Ornith. Club*, Vol. III., pp. 79, 80, April, 1878.

SALVIN, OSBERT.—Salvin on the Procellariidæ. Rowley's Ornithological Miscellany. Part IV. London, 1887.

... This paper is in two parts. The first is devoted to an examination of the unpublished "Banks' drawings," and the manuscripts of Dr. Solander, so far as they relate to the Petrels ... Mr. Salvin's second paper is a careful examination of the new species of Petrels obtained by Dr. H. H. Giglioli during the voyage of the Italian corvette "Magenta" round the world..... T. M. B., *Bull. Nutt. Ornith. Club*, Vol. II., pp. 69, 70, July, 1877.

SHARPE, R. BOWDLER.—Catalogue of the Birds of the British Museum. Vol. III. Catalogue of the Coliomorphæ, containing the families Corvidæ, Paradiseidæ, Oriolidæ, Dicruridæ, and Prionopidæ. By R. Bowdler Sharpe. 8vo., pp. xiii., 344, pll. xiv. 1877.

In the third volume Mr. Sharpe enters upon the great series of Passerine Birds The species here described by Mr. Sharpe number three hundred and sixty-seven.... We are sorry to see several instances of the use of the same name in a generic and specific sense for the same species....—J. A. A., *Bull. Nutt. Ornith. Club*, Vol. III., pp. 78, 79, April, 1878.

STREETS, THOMAS H.—Contributions to the Natural History of the Hawaiian and Fanning Islands and Lower California, made in connection with the United States North Pacific Surveying Expedition, 1873-75. By Thos. H. Streets, M.D., Passed Assistant Surgeon, U. S. Navy. *Bull. U. S. Nat. Mus.*, No. 7, 8vo., (Birds, pp. 9-33), Washington, 1877.

... includes notes on about fifty species of birds, of which rather more than one-half were collected on the coast of Lower California and adjoining portions of the Mexican coast. The author acknowledges his indebtedness to Dr. Elliott Coues, U. S. A., for the identification of the birds, and adds that he has "kindly furnished the notes accompanying that portion of the ornithological collection from the Californian Peninsula".... there are many valuable biographical and other notes on several hitherto little known species.—J. A. A., *Bull. Nutt. Ornith. Club*, Vol. III., pp. 80, 81, April, 1878.

WILLARD, S. L.—A List of the Birds of Central New York. Utica, N. Y., 1877. By S. L. Willard, Esq. 16 pp.

The author's remarks in the way of a prelude are thus briefly expressed: "The following is a complete list of the birds of Central New York, with notes on their abundance." This might lead one to expect a valuable contribution to our science, but a perusal of the "List" proves this supposition to be erroneous....—C. H. M., *Bull. Nutt. Ornith. Club*, Vol. III., pp. 83, 84, April, 1878.

1878.

Allen, J. A. A List of the Birds of Massachusetts, with Annotations, by J. A. Allen. *Bull. Essex Inst.*, Vol. X., pp. 3-37, April, 1878.

It is seldom that one meets with a local catalogue more thoroughly satisfactory in all essential respects than the present one ... this list presents the names of three hundred and sixteen species of ascertained occurrence in Massachusetts, not one of which can be challenged.... one hundred and thirty-five are marked as breeding within the State ... Thirty-five North American birds have been added to the Massachusetts list since 1867. T. M. B., *Bull. Nutt. Ornith. Club*, Vol. III., pp. 138-140, July, 1878.

Aughey, Samuel.—Notes on the Nature of the Food of the Birds of Nebraska. By Professor Samuel Aughey, of Lincoln, Neb. *First Ann. Rep. U. S. Ent. Comm. for the Year* 1877. Appendix II., pp. 13-62. 1878.

....The list numbers two hundred and fifty species, and hence includes a pretty large proportion of the birds that visit the State, and as the list relates ostensibly to only locust-eating species, our first feeling is one of surprise that it should be so large. Although Mr. Aughey's paper bears especially upon the subject of birds as grasshopper destroyers, it forms at the same time a valuable faunal list of the birds of Southern Nebraska, containing notes relating to the relative abundance and season of most of the species. J. A. A., *Bull. Nutt. Ornith. Club*, Vol. IV., pp. 110, 111, April, 1879.

Aughey, Samuel.—Some facts and considerations concerning the beneficial work of birds. By Professor Samuel Aughey, of Lincoln, Neb. *First Ann. Rep. U. S. Ent. Comm. for the Year* 1877, pp. 338-350. 1878.

....a special communication on the general subject of the usefulness of birds, with particular regard, however, to the locust question ....he concludes that even the majority of Raptorial birds should be protected ....He believes that sooner or later the protection of useful birds should become not only a national, but an international matter. .... —J. A. A., *Bull. Nutt. Ornith. Club*, Vol. IV., pp. 111, 112, April, 1879.

Brewer, T. M.—Notes on certain Species of New England Birds, with Additions to his Catalogue of the Birds of New England. By T. M. Brewer. *Proc. Boston Soc. Nat. Hist.*, Vol. XIX., pp. 301-309, April, 1878.

This paper adds twenty-one species to the "Catalogue of the Birds of New England," published by this author in 1875, and contains notes on twenty-seven other species of rare occurrence in New England. The whole number of "recognized forms" now admitted by him as having been taken in New England is three hundred and fifty-six .... J. A. A., *Bull. Nutt. Ornith. Club*, Vol. III., p. 185, October, 1878.

Bureau, Louis.—De la Mue du Bec et des Ornements Palpébraux du

Macareux arctique, Fratercula arctica (Lin.) Steph., après la saison des amours. Par le Docteur Louis Bureau. *Extrait du Bulletin de la Société Zoologique de France*, 1877. 8vo. Paris, 1878. Pp. 1–21, pll. iv., v.

The remarkable changes which the bill and eyelids of the Common Puffin undergo after the breeding season have been hitherto unknown The author's exposition of the matter reveals a phenomenon as yet unparalleled among birds The author concludes this remarkable paper with some pertinent and suggestive observations on other species of *Fratercula* and on *Lunda cirrhata*. Elliott Coues, *Bul. Nutt. Ornith. Club*, Vol. III., pp. 87 91, April 1878.

Cory, Charles B.—A Naturalist in the Magdalen Islands; giving a Description of the Islands, and List of the Birds taken there, with other Ornithological Notes. By Charles B. Cory. Illustrated from Sketches by the Author. Boston, 1878. Small 4to. Part II., Catalogue of Birds taken or observed in the Magdalen Islands, with Notes regarding those found breeding, etc., etc. Pp. 33–83.

In a sumptuous little quarto Mr. C. B. Cory has given an account of a summer trip to the Magdalen Islands in the Gulf of St. Lawrence. ....Part I. consists of a general account of the Islands . and directions how to reach the Magdalen group, etc. Part II. gives a list of one hundred and nine species observed and taken by the author .. The annotations relate mainly to the habits and relative abundance of the species ...—J. A. A., *Bull. Nutt. Ornith. Club*, Vol. IV., p. 171, July, 1879.

Coues, Elliott.—Birds of the Colorado Valley. A Repository of Scientific and Popular Information concerning North American Ornithology. By Elliott Coues. Part First. Passeres to Laniidæ. Bibliographical Appendix. Seventy illustrations (woodcuts). 8vo. Pp. xvi., 807. Washington: Government Printing Office, 1878. "Miscellaneous Publications, No. 11," of the United States Geological Survey of the Territories, F. V. Hayden, U. S. Geologist-in-Charge.

In point of completeness, mode of execution, and general usefulness, the bibliography here under notice far excels any natural history bibliography known to us, and deserves to rank with the best bibliographies of any department of literature, and may well serve as a model for future workers in similar fields.... As regards the general work, or the "Birds of the Colorado Valley" as a whole, no more important contribution to the subject of North American Ornithology than this promises to be has for a long time appeared, and none covering all points of the field here taken ;....—J. A. A., *Bull. Nutt. Ornith. Club*, Vol. IV., pp. 54–57, January, 1879.

Coues, Elliott.—Field Notes on Birds observed in Dakota and Montana along the Forty-ninth Parallel during the Seasons of 1873 and 1874. By Dr. Elliott Coues, U. S. A., late Surgeon and Naturalist

U. S. Northern Boundary Commission. *Bull. U. S. Geol. Survey of the Territories*, Vol IV., No. 3, pp. 515 661. July 29, 1878.

The observations relate mainly to the country ... from Pembina on the Red River to the Rocky Mountains ....a distance of about eight hundred and fifty miles. Dr. Coues in his preliminary remarks divides the country traversed into three regions, which he terms respectively the "Red River Region," the "Missouri Region," and the "Rocky Mountain Region" The physical and zoological characteristics of these regions are briefly detailed, to which is added a tabular enumeration of some of the more conspicuous birds of the three regions. Then follows a copiously annotated list of all the species observed .. J. A. A., *Bull. Nutt. Ornith. Club*, Vol. IV., pp. 49, 50, January, 1879.

JORDAN, DAVID STARR —Manual of the Vertebrates of the United States, including the District east of the Mississippi River, and north of North Carolina and Tennessee, exclusive of Marine Species. By David Starr Jordan, Ph.D., M.D., etc. Second Edition, revised and enlarged. Chicago: McClurg & Co., 1878. 12mo., pp. 407.

....the second edition has not only been to some extent "revised," but enlarged by the addition of upwards of fifty pages of new matter. ..The account of the fishes has been entirely re written ...—J. A A., *Bull. Nutt. Ornith. Club*, Vol. III., pp. 145, 146, July, 1878.

LAWRENCE, GEORGE N.—Descriptions of Seven New Species of Birds from the Island of St. Vincent, West Indies. By George N. Lawrence. *Ann. N. Y. Acad. Sci.*, Vol. I., pp. 146-152. Issued May-September, 1878.

LAWRENCE, GEORGE N.—Descriptions of Supposed New Species of Birds from the Islands of Grenada and Dominica, West Indies. By George N. Lawrence. *Ann. N. Y. Acad. Sci.*, Vol. I., pp. 160-163. Issued May-September, 1878.

MAYNARD, C J.—The Birds of Florida, with the Water and Game Birds of Eastern North America. By C. J Maynard. Illustrated. 4to. Part IV., pp. 89-112, and one Plate. C. J. Maynard & Co., Newtonville, Mass, 1878.

...is wholly devoted to the family *Fringillidæ*, of which fourteen species are described ..It is illustrated with a fine colored plate of the Ipswich or Pallid Sparrow (*Passerculus princeps*), representing the adult in spring. To original, somewhat detailed descriptions of the different phases of plumage of the various species treated the author adds short, very pleasantly written descriptions of their habits . J. A. A., *Bull. Nutt. Ornith. Club*, Vol. III., p. 145, July, 1878.

RIDGWAY, ROBERT.—Studies of the American Herodiones. Part I.—Synopsis of the American genera of Ardeidæ and Ciconiidæ; including descriptions of three new genera, and a monograph of the

American species of the genus Ardea. By Robert Ridgway. *Bull. U. S. Geol. and Geogr. Surv. of Terr.*, Vol. IV., pp. 219-251, February 5, 1878.

The first of a series of papers here begun deals mainly with the *Ardeidæ* and *Ciconiidæ* ....The *A. würdemanni* of Baird, which has been a puzzle to ornithologists for twenty years is considered to be the "blue phase" of *A. occidentalis* . - J. A. A., *Bull. Nutt. Ornith. Club*, Vol. III., pp. 182, 183, October, 1878.

Saunders, Howard.—On the Larinæ. By Howard Saunders. *Proc. Zoöl. Soc. London*, 1878, pp. 115-212.

The whole number of species recognized in this paper is forty-nine, of which number twenty may be counted as North American ... Mr. Saunders's paper evinces a remarkable success in disentangling the complicated web of European Gulls....and the service rendered by Mr. Saunders cannot fail to be appreciated by all who have experienced its need.—T. M. B., *Bull. Nutt. Ornith. Club*, Vol. III., pp 185 187, October, 1878.

Sennett, George B.—Notes on the Ornithology of the Lower Rio Grande of Texas, from Observations made during the Season of 1877. By George B. Sennett. Edited, with Annotations, by Dr. Elliott Coues, U. S. A. *Bull. U. S. Geol. and Geogr. Surv. of Terr.*, Vol. IV., pp. 1-66, February 5, 1878.

....on one hundred and fifty-one species of birds observed on the southern border of Texas....Mr. Sennett certainly collected under many annoyances, but intensely hot days....did not prevent his securing some five hundred birds, one of which is new to science, namely, Sennett's Warbler (*Parula nigrilora*). The paper is most carefully commentated by Dr. Coues ...—H. A. P., *Bull. Nutt. Ornith. Club*, Vol. III., pp. 144, 145, July, 1878.

Stevenson, H.—Adams's Notes on the Birds of Alaska. By H. Stevenson. *The Ibis*, 4th Series, Vol. II., pp. 420-442, Oct., 1878,

Some twenty-eight years ago (October, 1850) Mr. Edward Adams, a surgeon in the British navy ...was sent to the Redoubt of Michalaski, on the shores of Norton Sound, Alaska. He remained there until late in the following June, and made some very interesting and valuable notes on the birds of the region. His collections were given to the British Museum, to Mr. John Gould, and to the late Mr. G. R. Gray. The latter dedicated to him the *Colymbus adamsi* .. These early observations of Alaskan species... have intrinsic interest and are well worthy of attention.—T. M. B., *Bull. Nutt. Ornith. Club*, Vol. IV., pp. 52, 53, January, 1879.

Wilson, Alexander, and Bonaparte, Charles Lucian.—American Ornithology; or, The Natural History of the Birds of the United States. Illustrated with plates engraved from drawings from Nature. By Alexander Wilson and Charles Lucian Bonaparte. Popular edition. Philadelphia: Porter and Coates. Three volumes in one.

It claims to be an exact reproduction, minus the atlas of colored plates, of the $100, three-volume edition issued by the same firm some years ago. No one can help rejoicing at any effort to disseminate more widely an acquaintance with Alexander Wilson and his charming and painstaking work... But simply to reprint Wilson, even with Bonaparte added, at $7 50, pointing out none of the errors nor supplementing the shortcomings is, to say the least, utterly unnecessary to the advancement of the science.- E. I , *Bull. Nutt. Ornith. Club*, Vol. IV., pp 53, 54, January, 1879.

1879.

BELDING, L.—A Partial List of the Birds of Central California. By L. Belding, of Stockton. Edited by R. Ridgway. *Proc. U. S. Nat. Mus.*, Vol. I., April, 1879, pp. 388-449.

... It is based,... upon observations extending through about twenty years' residence in California, and upon collections made chiefly during the last two years, which have from time to time, been forwarded by Mr. Belding to the National Museum The number of species, exclusive of the wading and swimming birds, is 158 .. In respect to the designation of incipient species, Mr. Ridgway uniformly adopts the system advocated by him in his paper on the use of trinomials in zoölogical nomenclature in the present number of the Bulletin .. As already stated Mr. Ridgway was the first to adopt the system of pure trinomials, and we regret to note his divergence therefrom ...- J. A. A., *Bull Nutt. Ornith. Club*, Vol. IV., pp. 167 171, July, 1879.

BREWER, T. M.—Some Additional Notes upon Birds observed in New England, with the Names of Five Species not included in his Previous Lists of New England Birds. By T. M. Brewer. *Proc. Boston Soc. Nat. Hist.*, Vol. XX., pp. 263 277. Published December, 1879.

....It forms a second supplement to his "Catalogue of the Birds of New England," published in 1875, and adds five species to the number previously recognized by him as New England birds, raising the whole number to 361 . These "Notes" form a convenient and connected record of recent discoveries in relation to many of the rarer New England birds, and add more or less that is new respecting some of them.—J. A. A., *Bull. Nutt. Ornith. Club*, Vol. V., pp. 108, 109, April, 1880.

COUES, ELLIOTT.—On the Present Status of *Passer domesticus* in America, with Special Reference to the Western States and Territories. By Dr. Elliott Coues, U. S. A. *Bull. U. S. Geol. and Geogr. Surv. of Terr.*, Vol. V., pp. 175 193, Sept. 6, 1879.

....a partial bibliography of, what is commonly termed the "Sparrow-War in America" in which are given the titles of most of the papers relating to this troublesome question, usually with a short digest of the papers mentioned J. A. A., *Bull. Nutt. Ornith Club*, Vol. V., p. 41, January, 1880.

COUES, ELLIOTT.—Second Instalment of American Ornithological Bibliography. By Dr. Elliott Coues, U. S. A. *Bull. U. S. Geol. and Geogr. Surv. of Terr.*, Vol. V., pp. 239 330, September 6, 1879.

....This part gives the titles of "Faunal Publications" relating to Central and South America, or that portion of America forming the so-called "Neotropical Region." ....Beginning with Marcgrave in 1648, the list of titles is brought down to include most of those which appeared in the first half of the year 1879. ....Of the laborious research and care displayed in the preparation of this work, too great praise can scarcely be accorded.- J. A. A., *Bull. Nutt. Ornith. Club*, Vol. V., pp. 40, 41, January, 1880.

DARTT, MARY.—On the Plains and among the Peaks; or, How Mrs. Maxwell made her Natural History Collection. By Mary Dartt. Philadelphia: Claxton, Remsen, and Haffelfinger, 624, 626, 628 Market Street, 1879. 8vo., pp. 237.

Among the many wonderful "exhibits" at the recent Centennial Exposition in Philadelphia, few things attracted such general attention or created more surprise... than Mrs. M. A. Maxwell's collection of the animals of Colorado. This little book before us, devoted mainly to a very intelligent and pleasantly written account of how Mrs. Maxwell's work was accomplished, was prepared by a sister of that lady-naturalist. The main text of the work is intended for the general public. ; but in an "Appendix" of twenty pages are given annotated lists of the mammals and birds represented in the collection, the former by Dr. Coues and the latter by Mr. Ridgway....The list of birds numbers 234 species The annotations relate mainly to an enumeration of the specimens represented, but occasionally to facts of distribution and locality of occurrence.- J. A. A., *Bull. Nutt. Ornith. Club*, Vol. IV., pp. 113, 114, April, 1879.

ELLIOT, DANIEL GIRAUD.—A Classification and Synopsis of the Trochilidæ. By Daniel Giraud Elliot, F.R.S.E., etc. Washington City: Published by the Smithsonian Institution. March, 1879. 4to., pp. xii., 277, figg. 127 (wood-cuts in the the text).

....Mr. Elliot's concise and comprehensive "Synopsis" ...forms a most welcome aid to the student of this intricate group Four hundred and twenty-six species are admitted as valid, distributed among one hundred and twenty genera The leading characters of very nearly all the genera are represented by outline figures of the head, wing, and tail, and the species are described in sufficient detail for their easy recognition .. The work closes with an appendix, giving an analytical key to the genera, and separate indexes to the generic and specific names... It will doubtless form a reference work for the group, not to be soon superseded, either in point of completeness or of usefulness.—J. A. A, *Bull. Nutt Ornith. Club*, Vol. IV., pp. 230-232, October. 1879.

GIBBS, MORRIS.—Annotated List of the Birds of Michigan. By Dr. Morris Gibbs. *Bull. of the U. S. Geol. and Geogr. Surv. of Terr.*, Vol. V., No. 3, pp. 481-497, November 30, 1879.

Although several prior lists of the birds of Michigan have appeared, the present one is a welcome addition to our knowledge of the ornithology of that State. Mr. Gibbs's list enumerates 310 species and subspecies, and contains briet notes on their relative abundance, breeding, times of migration, etc . . Although mainly based on the observations of the author, he expresses his indebtedness to other sources of information ...—J. A. A., *Bull. Nutt. Ornith. Club*, Vol. V., p. 110, April, 1880

HALLOCK, CHARLES.—The Sportman's Gazetteer and General Guide. Fifth edition. By Charles Hallock.

....This book has become a recognized authority on all subjects of which it treats, having been already republished in England, France, and Germany.... The ornithological portions were, we believe, prepared by Mr. George B. Grinnell.—J. A. A., *Bull. Nutt. Ornith. Club*, Vol. IV., p. 175, July, 1879.

HARVIE-BROWN, JOHN A.—Ornithological Journal of the Winter of 1878-79, with Collected Notes regarding its Effects upon Animal Life, including Remarks on the Migration of Birds in the Autumn of 1878, and the Spring of 1879. By Mr. John A. Harvie-Brown, F.Z.S., M.B.O.U. *Proc. Nat. Hist. Soc. Glasgow*, 1879, pp. 123-190.

....The winter of 1878-79 proved of unusual severity, and its effect upon animal life, and especially upon bird life attracted the attention of many careful observers, Mr. Harvie-Brown giving a list of more than a dozen published papers relating to the subject. These with his own observations and the collected notes of his many correspondents, form the basis of the paper above cited,... nearly fifty pages being devoted to birds. J. A. A., *Bull. Nutt. Ornith. Club*, Vol. V., pp. 233, 234, October, 1880.

HARVIE-BROWN, JOHN A.—The Capercaillie in Scotland. By J. A. Harvie-Brown, F.Z.S., Member of the British Ornithologists' Union, etc. Edinburgh: David Douglas, 1879, 8vo., pp. i-xv, 1-155, map and pll.

....Mr. Harvie-Brown treats the general subject of the Capercaillie in Scotland exhaustively. Beginning with such prehistoric evidence as is afforded by the bone-caves and Kitchen-middens.... He then presents its history prior to extinction, followed by that of its restoration, and a detailed account of its increase and extension, illustrated by a map. He later discusses the influences which govern its increase, its relation to the decrease of Black Game, its damage to forests and grain, etc. Of special interest also are his chapters on the derivation, significance, and proper orthography of the word Capercaillie. In short, every point of the subject is elaborated with the utmost thoroughness, the work forming a model of its class.—J. A. A., *Bull. Nutt. Ornith. Club*, Vol. V., pp. 110, 111, April, 1880.

INGERSOLL, ERNEST.—Nests and Eggs of American Birds. By Ernest Ingersoll. S. E. Cassino, Naturalists' Agency, Salem, Mass. (No date.) Large 8vo. Part I., pp. 1-24, pll. i, ii., March, 1879.

....treats of ten species of Thrushes, and gives illustrations of their eggs. The text includes, not only descriptions of the nests and eggs of the species treated, but a full and pleasantly written account of their habits and breeding range ... We wish that we could speak in terms of equal commendation of the chromo-lithographic plates, which are sadly defective in point of faithfulness to nature and in artistic execution.—J. A. A., *Bull. Nutt. Ornith. Club*, Vol IV., p. 172, July, 1879.

——— Part II., pp. 25-48, pll. iii, iv., published August, 1879.

....we regret to perceive that the parts continue to appear without dating, or any indications whatever of the time of their publication;

and that textual references to the figures of the plates are still insufficiently explicit ... Mr. Ingersoll has his subject well in hand now; he confines himself strictly to the announced scope of the treatise, and holds his subject fairly abreast of the information we have acquired respecting it.—E. C., *Bull. Nutt. Ornith. Club*, Vol. V., pp. 38, 39, January, 1880.

———Part III., pp. 49-72, pll. v., vi., published October, 1879.

Krider, John.—Forty Years' Notes of a Field Ornithologist, by John Krider, Member of the Philadelphia Academy of Natural Sciences and author of Krider's Sporting Anecdotes, Philadelphia. Giving a description of all birds killed and prepared by him. Philadelphia, 1879, 8vo., pp. i-xi., 1-84.

....Mr. Krider has "endeavored to describe and give the history of only those species of birds of the United States" which he has "collected and mounted," and whose nests have come under his personal observation .. But a casual glance through the pages of his work is enough to show that these opportunities have been sadly neglected... In short, it is only too evident that Mr. Krider's "Notes" are the offspring of a fading memory rather than the carefully kept data of a systematic worker....Of the literary execution of the present work we can say nothing favorable....—W. B., *Bull. Nutt. Ornith. Club*, Vol. VII., pp. 49, 50, January, 1882.

Kumlien, Ludwig.—Contributions to the Natural History of Arctic America, made in Connection with the Howgate Polar Expedition, 1877–78. By Ludwig Kumlien, Naturalist of the Expedition. *Bull. U. S. Nat. Mus.*, No. 15, 1879. Birds, pp. 69-105.

...Of the 84 species noted, seven or eight relate to localities not Arctic, being species that visited the ship while off Newfoundland and neighboring points. Of the remainder only about twenty are land birds. The notes respecting many of the species are quite extended, and embrace many points of interest....—J. A. A., *Bull. Nutt. Ornith. Club*, Vol. V., pp. 109, 110, April, 1880.

Lawrence, George N.—Catalogue of the Birds of Dominica, from Collections made for the Smithsonian Institution by Frederick A. Ober, together with his Notes and Observations. By George N. Lawrence. *Proc. U. S. Nat. Mus.*, Vol. I., 1879, pp. 48-69.

Lawrence, George N.—Catalogue of the Birds of St. Vincent, from Collections made by Mr. Frederick A. Ober, under the Directions of the Smithsonian Institution, with his Notes thereon. By George N. Lawrence. *Proc. U. S. Nat. Mus.*, Vol. I., 1879, pp. 185–198.

Lawrence, George N.—Catalogue of the Birds of Antigua and Barbuda, from Collections made for the Smithsonian Institution, by Mr. Fred. A. Ober, with his Observations. By George N. Lawrence. *Proc. U. S. Nat. Mus.*, Vol. I., 1879, pp. 232-242.

LAWRENCE, GEORGE N.—Catalogue of the Birds of Grenada, from a Collection made by Mr. Fred. A. Ober for the Smithsonian Institution, including others seen by him, but not obtained. By George N. Lawrence. *Proc. U. S. Nat. Mus.*, Vol. I., 1879, pp. 265–278.

LAWRENCE, GEORGE N.—Catalogue of the Birds collected in Martinique by Mr. Fred. A. Ober for the Smithsonian Institution. By George N. Lawrence. *Proc. U. S. Nat. Mus.*, Vol. I., 1879, pp. 349–360.

LAWRENCE, GEORGE N.—Catalogue of a Collection of Birds obtained in Guadeloupe for the Smithsonian Institution, by Mr. Fred. A. Ober. By George N. Lawrence. *Proc. U. S. Nat. Mus.*, Vol. I., 1879, pp. 449–462.

LAWRENCE, GEORGE N.—A General Catalogue of the Birds noted from the Islands of the Lesser Antilles visited by Mr. Fred. A. Ober; with a Table showing their Distribution, and those found in the United States. By George N. Lawrence. *Proc. U. S. Nat. Mus.*, Vol. I., 1879, pp. 486–488.

....he has concluded his series of reports upon Mr. Ober's collections, made at various points of the Antillean chain (see above). The birds reported from Antigua and Barbuda number respectively 42 and 39 species, of which one....from Antigua, is described as new. ....The list of birds from the island of Grenada numbers 51 species. . The birds reported from Martinique number 40 species .. The Guadeloupe species number 45 ... —J. A. A., *Bull. Nutt. Ornith. Club*, Vol. IV., pp. 228–230, October, 1879.

LANGDON, FRANK W.—A Revised List of Cincinnati Birds. By Frank W. Langdon. *Journ. Cincinnati Soc. Nat. Hist.*, Vol. I., No. 4, January, 1879, pp. 167–193.

....The 256 identified species are of the following categories: Constant residents, 27; summer residents, 62; winter visitants, 10; regular migrants, 82; irregular migrants, 37; casual visitants, 31; species that have disappeared within forty years, 7 . It is a very good piece of work, based in greatest part on original personal observations, very carefully elaborated, with attention not only to the material facts presented, but to those niceties of workmanship which are too often neglected.... We are glad to see, especially among our younger writers on ornithology, evidence of increased attention to details of execution.... an article may be made a contribution to letters as well as to science. It is even worth while to spell correctly. E. C., *Bull. Nutt. Ornith. Club*, Vol. IV., pp. 112, 113, April, 1879.

MEARNS, EDGAR A.—A List of the Birds of the Hudson Highlands, with Annotations. By Edgar A. Mearns. *Bull. Essex Institute*, Vol. X., pp. 166–179 (Introduction and *Turdus migratorius* to *Parus atricapillus*, inclusive), October–December, 1878.

MEARNS, EDGAR A.—A List of the Birds of the Hudson Highlands, with Annotations. By Edgar A. Mearns. *Bull. Essex Institute*, Vol. XI., pp. 43-52 (*Sitta carolinensis* to *Dendrœca cærulescens*), January-March, 1879.

———*Bull. Essex Institute*, Vol. XI., pp. 154-168 (*Dendrœca cærulea* to *Myiodioctes mitrata*), July-September, 1879.

———*Bull. Essex Institute*, Vol. XI., pp. 189-204 (*M. canadensis* to *Loxia curvirostra*,), October-December, 1879.

The first part . appeared early in 1879, and three later instalments carry the list through the genus *Loxia*.... while the writer draws mainly from his own experience, he occasionally indulges in quotations from other authors, his notices of some of the species amounting to nearly complete biographies ...Two important features of the paper are the dates of arrival and departure,....The future instalments of Mr. Mearns's highly praiseworthy memoir may well be anticipated with interest.—J. A. A., *Bull. Nutt. Ornith. Club*, Vol. V., p. 175, July, 1880.

MCCHESNEY, CHARLES E.—Notes on the Birds of Fort Sisseton, Dakota Territory. By Chas. E. McChesney, Acting Assistant Surgeon, U. S. A. *Bulletin U. S. Geol. and Geogr. Surv. Terr.*, Vol. V., pp. 71-104, February 28, 1879.

....form a valuable contribution to the ornithology of a little known portion of the Northwest, namely, the elevated plateau in Dakota, known as the "*Coteau des Prairies.*" .. The "Notes" are based on an experience of three years in the neighborhood of Fort Sisseton, and record 157 species, respecting most of which there are copious and interesting annotations....Dr. McChesney's report was transmitted to Dr. Coues for publication, and appears to have had the benefit of his revision....—J. A. A., *Bull. Nutt. Ornith. Club*, Vol. V., pp. 42, 43, January, 1880.

MCCHESNEY, CHARLES E.—Report on the Mammals and Birds of the General Region of the Big Horn River and Mountains of Montana Territory. By Charles E. McChesney, U. S. A. Being Appendix SS 3 of the Report of the Chief of Engineers for 1879.

....proves an interesting addition to the faunal records of the West ...The list of 100 species of birds is the result of less than a month's investigation—from August 15 onward—....The notes, though brief, are usually sufficient to indicate the occurrence of each species, and, as in the greater number of cases they result directly from the author's own observations, they carry with them the value of perfect authenticity... contains the name of not a single exclusively Eastern species....—H. W. H., *Bull. Nutt. Ornith. Club*, Vol. V., pp. 107, 108, April, 1880.

MERRILL, JAMES C.—Notes on the Ornithology of Southern Texas. Being a List of Birds observed in the Vicinity of Fort Brown, Texas, from February, 1876, to June, 1878. By James C. Merrill, Assistant Surgeon U. S. Army. *Proc. U. S. Nat. Mus.*, Vol. I., 1879, pp. 118-173.

....Two hundred and fifty-two species and varieties are given in all, and the character of their presence is in most cases satisfactorily detined ...the nests, eggs, and breeding habits of Texan birds receive the larger share of attention, and much of the matter pertaining thereto is as valuable as it is now ...Numerous notes by Mr. Ridgway and Dr. Brewer occur throughout the paper and greatly enhance its value ...In a few details of arrangement the paper is open to adverse criticism ... Altogether, however, the paper is a most excellent one, and its contents supply a fund of information the lack of which has been long felt. - W. B., *Bull. Nutt. Ornith. Club*, Vol. IV., pp. 50-52, January, 1879.

RATHBUN, FRANK R.—A Revised List of Birds of Central New York. Based on the Observations of Frank R. Rathbun, H. Gilbert Fowler, Frank S. Wright, Samuel F. Rathbun, in the Counties of Cayuga, Onondaga, Seneca, Wayne, and Yates. Collated and prepared for Publication by Frank R. Rathbun. Auburn, N. Y.: Daily Advertiser and Weekly Journal Book and Job Printing House, April 17, 1879.

....in the present "Revised List" are enumerated 236, showing an addition of 46 species ...In conclusion, it is but just to say that "The Ornithological Four" have in their "Revised List of Birds of Central New York," not only done themselves great credit, but have made a contribution to our science which must long remain authority concerning the region of which it treats. I consider it the best list of the birds of any part of this State that has appeared for many years.- C. H. M., *Bull. Nutt. Ornith. Club*, Vol. IV., pp. 172-175, July, 1879.

RIDGWAY, ROBERT.—On a new Humming-bird (Atthis ellioti) from Guatemala. By Robert Ridgway. *Proc. U. S. Nat. Mus.*, Vol. I., 1879, pp. 8–10.

RIDGWAY, ROBERT.—A Review of the American Species of the Genus Scops, Savigny. By Robert Ridgway. *Proc. U. S. Nat. Mus.*, Vol. I., 1879, pp. 85-117. Author's separates issued August 6, 1878.

RIDGWAY, ROBERT.—Description of Several New Species and Geographical Races of Birds Contained in the Collection of the United States National Museum. By Robert Ridgway. *Proc. U. S. Nat. Mus.*, Vol. I., 1879, pp. 247–252. Author's separates issued December 10, 1878.

RIDGWAY, ROBERT.—Descriptions of Two New Species of Birds from Costa Rica, and Notes on other Rare Species from that Country. By Robert Ridgway. *Proc. U. S. Nat. Mus.*, Vol. I., 1879, pp. 252–255. Author's separates issued December 10, 1878.

RIDGWAY, ROBERT.—Descriptions of New Species and Races of American Birds, including a Synopsis of the Genus Tyrannus, Cuvier. By Robert Ridgway. *Proc. U. S. Nat. Mus.*, Vol. I., 1879, pp. 466–486. Author's separates issued April 25, 1879.

These (five) papers all notably evince Mr. Ridgway's well-known acuteness of discrimination and critical care in description and diagnosis.... -J. A. A., *Bull. Nutt. Ornith. Club*, Vol. V., pp. 41, 42, January, 1880.

ROOSEVELT, THEODORE.—Notes on some of the Birds of Oyster Bay, Long Island. By Theodore Roosevelt. 8vo., 1 p. March, 1879.

This is a brochure of a single leaf, containing notes on seventeen species, observed at the above-named locality, by Mr. Theodore Roosevelt ...—J. A. A., *Bull. Nutt. Ornith. Club*, Vol. IV., p. 171, July, 1879.

SENNETT, GEORGE B.—Further Notes on the Ornithology of the Lower Rio Grande of Texas, from Observations made during the Spring of 1878. By George B. Sennett. Edited, with Annotations, by Dr. Elliott Coues, U. S. A. *Bull. U. S. Geol. and Geogr. Surv. of Terr.*, Vol. V., No. 3, pp. 371–440, November 30, 1879.

The report of Mr. Sennett's three months' work (in April, May, and June) in 1878, near Hidalgo, Texas, adds greatly to our knowledge of the life-histories of many species of which we previously knew but little. ....In addition to the notes on the habits of the birds observed, which in the case of the less known species amounts in some instances to full biographies, the author presents us with extended tables of measurements, gives detailed descriptions of nest and eggs, and occasionally discusses points of relationship and nomenclature....The "Notes" relate to 168 species, and altogether form one of the most valuable of the many recent contributions to local ornithology.—J. A. A., *Bull. Nutt. Ornith. Club*, Vol. V., p. 111, April, 1880.

SHARPE, R. BOWDLER.—Catalogue of the Birds in the British Museum. Vol. IV. Catalogue of the Passeriformes, or Perching Birds, in the British Museum. Cichlomorphæ: Part I., containing the families Campophagidæ and Muscicapidæ. By R. Bowdler Sharpe. London, 1879. 8vo., pp. xvi., 494, pll. xiv.

Of the *Campophagidæ* 148 species are described, of the *Muscicapidæ* 391. In style of treatment and general character this volume is similar to the earlier ones....—J. A. A., *Bull. Nutt. Ornith. Club*, Vol. VIII., p. 99, April, 1883.

VOGT, M. C.—L'Archæopteryx macroura.—Un intermédiaire entre les oiseaux et les reptiles. Par M. C. Vogt. La Revue Scientifique, 2e Séries, 9e Année, No. 11, 13 Sept. 1879, pp. 241–248, figg. 18–21.

This specimen was found by M. Haeberlein in the same slates as the first....From what Professor Vogt has discovered by a cursory examination there can be no doubt that much of great interest will be learned when this fossil is properly worked out from the matrix.—J. A. Jeffries, *Bull. Nutt. Ornith. Club*, Vol. VI., pp. 107–109, April, 1881.

1880.

BELL, ROBERT.—List of Birds from the Region between Norway House and Forts Churchill and York. [By Robert Bell.] *Geological Survey of Canada.* Report of Progress for 1878–79 (1880). IV., Appendix vi., pp. 676–706.

....an annotated list of 55 species, of much interest from the localities of observation.....—J. A. A., *The Auk*, Vol. II., p. 209, April, 1885.

BRAYTON, ALEMBERT W.—A Catalogue of the Birds of Indiana, with Keys and Descriptions of the Groups of greatest Interest to the Horticulturist. By Alembert W. Brayton, B.S., M.D. *Transactions of the Indiana Horticultural Society* for 1879, pp. 89–166. Indianapolis, 1880.

... is intended as a "practical hand-book" of the Birds of Indiana, and seems well calculated to meet this requirement. It is avowedly a compilation....we note little in Dr. Brayton's paper that is new to ornithologists, but much that is given from good authorities. Short notes are added relative to the abundance, habits, and season of occurrence of the 306 species enumerated . The paper closes with an index to the names of the genera and higher groups, with their derivations, a "glossary" of the specific names, and an index of English names. ...—J. A. A., *Bull. Nutt. Ornith. Club*, Vol. V., pp. 174, 175, July, 1880.

BREWER, T. M.—Notes on the Nests and Eggs of the Eight North American Species of Empidonaces. By T. M. Brewer. *Proc. U. S. Nat. Mus.*, Vol. II., 1880, pp. 1–10. Author's separates issued April 29, 1879.

... Following the measurements and descriptions of the nests and eggs of these eight species are several pages devoted to a consideration of the nests and eggs of *E. flaviventris* .. —J. A. A., *Bull. Nutt. Ornith. Club*, Vol. IV., p. 232, October, 1879.

COOPER, J. G.—On the Migrations and Nesting Habits of West-Coast Birds. By J. G. Cooper, M.D. *Proc. U. S. Nat. Mus.*, Vol. II., 1880, pp. 241–251. Author's separates issued Jan. 20, 1880.

....Dr. Cooper has tabulated a large amount of valuable information respecting the times of arrival, departure, and nesting of many of the common West-Coast land birds, based mainly on his own observations ....The number of species tabluated is 73... Dr. Cooper has here begun a good work in a praiseworthy way,.... J. A. A., *Bull. Nutt. Ornith. Club*, Vol. V., p. 232, October, 1880.

CORY, CHARLES B.—Birds of the Bahama Islands; containing many Birds new to the Islands, and a Number of undescribed Winter Plumages of North American Birds. By Charles B. Cory, Author of "A Naturalist in the Magdalen Islands," etc. Illustrated. Boston: Published by the Author, 8 Arlington Street, Boston. 1880. 4to., pp. 250, with 8 colored plates.

....forms a valuable addition to our knowledge of the birds of these islands. Of the 149 species recorded, all but about 30 were met with by Mr. Cory. ...In addition to the short descriptions of the species, the relative abundance and distribution of the species is noted, to which is frequently added a short account of their habits .. J. A. A., *Bull. Nutt. Ornith. Club*, Vol. V., p 107, April, 1880.

COUES, ELLIOTT.—Third Instalment of American Ornithological Bibliography. By Dr. Elliott Coues, U. S. A. *Bull. U. S. Geol.*

and Geogr. Surv. of Terr., Vol. V., No. 4, 1879, pp. 521-1,066. Published Sept. 30, 1880.

...is by far the largest of the three,....and completes his "Bibliography of Ornithology so far as America is concerned"... The present third instalment consists of a selection of titles belonging to the "systematic" department....In reference to the character of the work, it is enough to say that it is fully up to the high standard of excellence of the previous instalments .. Its utility no working ornithologist can fail to highly appreciate, while it will form an enduring monument to the author's patience, industry, and thoroughness of research.—J. A. A., *Bull. Nutt Ornith. Club*, Vol. VI., pp. 44-46, January, 1881.

COUES, ELLIOTT.—Fourth Instalment of Ornithological Bibliography: being a list of Faunal Publications relating to British Birds. By Dr. Elliott Coues, U. S. A. *Proc. U. S. Nat. Mus.*, Vol. II., 1880, pp. 359-476. Published May 31, 1880.

This "Fourth Instalment" is of the same character as the first two, and attempts to do for British Birds what those did for American Birds ...As it is, being accurate as far as it goes, it will prove of great usefulness, and is entitled to the cordial welcome it will doubtless receive. —J. A. A., *Bull. Nutt. Ornith. Club*, Vol. VI., p. 46, January, 1881.

FORBES, S. A.—Studies of the Food of Birds, Insects, and Fishes, made at the Illinois State Laboratory of Natural History, at Normal, Illinois. By S. A. Forbes. *Illinois State Laboratory of Natural History Bulletin*, No. 3, November, 1880, 8vo., pp. 1-160.

.. a further report of his studies, about seventy pages of which relate to birds....The species of birds investigated are, as before, the Thrushes and the Bluebird ...—J. A. A., *Bull. Nutt. Ornith. Club*, Vol. VI., p. 110, April, 1881.

FREKE, PERCY EVANS.—A Comparative Catalogue of Birds found in Europe and North America. By Percy Evans Freke. Dublin, 1880. 8vo., pp. 44. *From the Scientific Proceedings of the Royal Dublin Society.*

....forms an important contribution to geographical ornithology, About 225 North American species are enumerated....Of about 100 species that may be considered as merely stragglers from one continent to the other, fully four-fifths are North American....Despite a few typopographical errors....the paper gives evidence of careful preparation and admirably fills a long-standing gap in ornithological literature.— J. A. A., *Bull. Nutt. Ornith. Club*, Vol. V., pp. 173, 174, July, 1880.

GENTRY, THOMAS G.—Illustrations of Nests and Eggs of Birds of the United States, with Text, by Thos. G. Gentry. Philadelphia . J. A. Wagenseller, Publisher, No. 23 North Sixth Street. Copyright by J. A. Wagenseller, 1881. 4to., parts 1-25, pp. 1-300. 54 colored chromo-lithographs and chromo-portrait frontispiece of the author. 1880-82.

....the plates were executed by Mr. Edwin Sheppard, "subject to the suggestions and dictations of the author." The title is misleading for instead of treating of all the species found in the United States it deals with but fifty .. The typography and press work are good, but the plates fall far short of deserving the same praise....of most of the plates....the perspective is very bad....and .. nearly all have the appearance of cheap chromo-lithographs....the work does not contain anything approaching a complete "detailed account of the habits" of a single species ...instead of becoming an authority....Mr. Gentry's book on nests and eggs must inevitably find its level alongside such unreliable and worthless productions as Jasper's "Birds of North America"....—C. H. M., *Bull. Nutt. Ornith. Club*, Vol. VII., pp. 246-248, October, 1882.

GREGG, W. H.—Revised Catalogue of the Birds of Chemung County, New York. By W. H. Gregg, M.D., Elmira, N. Y.: O. H. Wheeler. 1880.

....we have a list of the birds of a locality to which little attention has been paid by ornithologists. The list of which this is a revision was issued ten years ago ...In all. 217 species are enumerated... A few lines of notes accompany each name ...—E. I., *Bull. Nutt. Ornith. Club*, Vol. V., p. 173, July, 1880.

HARVIE-BROWN, J. A.—The Capercaillie in Scotland. By J. A. Harvie-Brown, F.R.S. *Scottish Naturalist*, July, 1880.

....Mr. Harvie-Brown published last year an exhaustive little work on the Capercaillie in Scotland. The present paper is a continuation of the Appendix of that work, giving an account of its extension in 1879, with a few additional references to early records of its presence in Scotland and Wales.—J. A. A., *Bull. Nutt. Ornith. Club*, Vol. VI., p. 46, January, 1881.

HARVIE-BROWN, JOHN A., and CORDEAUX, JOHN.—Report on the Migration of Birds in the Autumn of 1879. By John A. Harvie-Brown and John Cordeaux. *Zoölogist*, May, 1880, pp. 161-204.

... two well-known British ornithologists, have set themselves seriously at work in the matter of collecting exact data respecting the movements of birds during their migrations along the coasts of Great Britain....Observations made at other points are incidentally incorporated, including Herr Gätke's report from Heligoland. The work so earnestly begun....should be a stimulus to concurrent action on the part of others, and nowhere are the conditions more favorable for systematic work than in the United States.—J. A. A., *Bull. Nutt. Ornith. Club*, Vol. V., pp. 175-177, July, 1880.

HARVIE-BROWN, JOHN A.—Second Report on Scottish Ornithology—October 1, 1879, to September 30, 1880. Compiled by Mr. John A. Harvie-Brown, F.R.S.E., etc. *Proc. Nat. Hist. Soc. of Glasgow*, Vol. IV., Part II., April, 1880, pp. 291-326.

....The report gives a "Journal of the Winter of 1879-80" .. the report gives observations on some 65 to 70 species. The report abounds with especially suggestive observations in relation to little understood points of bird-life....—J. A. A., *Bull. Nutt. Ornith. Club*, Vol. VI., p. 174, July, 1881.

HENSHAW, H. W.—Ornithological Report upon Collections made in Portions of California, Nevada, and Oregon. By H. W. Henshaw. *Annual Report of the U. S. Geogr. Surveys west of the Hundredth Meridian, for* 1879. Appendix L of the Report of the Chief of Engineers, February, 1880, pp. 282–335.

Mr. H. W. Henshaw's "Ornithological Report" for the field seasons of 1877 and 1878 is much more than a record of field observations for the seasons named, treating as it does most ably, though briefly, of the relationships of the members of several of the most puzzling groups of North American birds. In addition to having access to a large amount of material, much of which the author collected himself, he is able to bring to bear upon the questions at issue an intimate knowledge of the birds in life, and of the varying conditions of environment which surround the forms treated ... In relation to the habits of the species mentioned, the Report contains much that is new, ... -J. A. A., *Bull. Nutt. Ornith. Club*, Vol. V., pp. 105 107, April, 1880.

LANGDON, FRANK W.—Ornithological Field Notes, with five Additions to the Cincinnati Avian Fauna. By Frank W. Langdon. *Journ. Cincinnati Soc. Nat. Hist.*, July, 1880, pp. 121–127, 1 pl.

These notes ... virtually form a supplement to the same author's excellent "Revised List of Cincinnati Birds" published in 1879.... They relate to 40 species . Among the points of special interest are the capture of two specimens (male and female) of Kirtland's Warbler (*Dendrœca kirtlandi*) near Cleveland, May 4 and 12, 1880....The paper is preceded by Dr. Langdon's description of a new species of *Helminthophaga*....—J. A. A., *Bull. Nutt. Ornith. Club*, Vol. V., pp. 232, 233, October, 1880.

MARSH, OTHNIEL CHARLES.—Odontornithes: a Monograph on the Extinct Toothed Birds of North America; with thirty-four Plates and forty Wood-cuts. By Othniel Charles Marsh, Professor of Palæontology in Yale College. 1 vol. 4to. Pp. i-x., 1-201, figg. 1–40, pll. i-xxxiv., each with 1 explanatory leaf. Forming Vol. VII. of the Reports of the Survey of the 40th Parallel.

....It is the first of a series of monographs designed to make known to science the extinct vertebrate life of North America, in the investigation of which the author has passed the last ten years. It is unquestionably the most magnificent contribution ever made to our knowledge of extinct birds ... It is safe to say that no single memoir on fossil birds hitherto published can be compared with this in accuracy of detail, in beauty of illustration, and in value of results attained ... The present volume is based on the remains of more than one hundred different individuals of the *Odontornithes* procured in the Cretaceous deposits of the West during the last ten years ...The work of Professor Marsh, as a whole, is an unmeasured advance upon all previously obtained knowledge of Cretaceous birds. The present volume is divided into two parts, the first treating of *Hesperornis*, the second of *Ichthyornis* and *Apatornis*, the entire skeleton of typical species being described with elaborate detail, and figured in the most perfect manner .. The Appendix presents a synopsis of the nine genera and twenty species of American Cretaceous Birds....—E. C., *Bull. Nutt. Ornith. Club*, Vol. V., pp. 234–236, October, 1880.

MAYNARD, C. J.—The Birds of Florida, with the Water and Game Birds of Eastern North America. By C. J. Maynard. Illustrated. Published by C. J. Maynard & Co., Newtonville, Mass.

....the eighth part has just been received The text is by far the most satisfactory part of the work, and contains much of interest, though, perhaps, too much space is given to the habits of some species as observed in New England and elsewhere ...Certain changes are made in nomenclature and classification, notably raising the Kingfishers and Nighthawks to the rank of orders ...Plates i., ii., iii., and xii. are passable, ...but the others are extremely poor, Plate vii., in Part vi., has figures of sixty-six eggs of sixty-four species.... J. C. M., *Bull. Nutt. Ornith. Club*, Vol. IV., pp. 114, 115, April, 1879.

MAYNARD, C. J.—The Birds of Eastern North America, with original Descriptions of all the Species which occur east of the Mississippi River between the Arctic Circle and the Gulf of Mexico, with full Notes upon their Habits. By C. J. Maynard. Containing thirty Plates drawn on Stone by the Author. C. J. Maynard & Co., Newtonville, Mass. 4to. (Thirteen Parts issued.)

See above, same work under another title.

... the peninsula (of Florida) has never received so much attention at the hands of any one ornithologist, not excepting Audubon, as from Mr. Maynard. It is a matter for regret that the later plans of the work had not been its original one. Had such been the case, the author would have been spared the necessity--if indeed it be a necessity—of repeating verbatim in the "Birds of Eastern North America" many pages of descriptive matter and biography which appeared in the "Birds of Florida"....In his classification Mr. Maynard has departed in many particulars from beaten paths, the basis for most of his changes being anatomical ...It is evident that the "Birds of Eastern North America" was written more with a view of striking the popular taste than as a hand-book for the systematic ornithologist, .. In conclusion, we may be permitted to express the feeling that the portions of the work now before us do not by any means represent the author's best efforts, and that in certain particulars, but especially as regards the plates, he is capable of placing the work on a far higher plane than can at present be accorded it. H. W. H., *Bull. Nutt. Ornith. Club*, Vol. V., pp. 170 173, July, 1880.

MEARNS, EDGAR A.—A List of the Birds of the Hudson Highlands, with annotations. By Edgar A. Mearns. *Bull. Essex. Institute*, Vol. XII., pp. 11-25 (*Ægiothus linaria* to *Quiscalus purpureus*), January–June, 1880.

——— *Bull. Essex. Institute*, Vol. XII., pp. 109 128 (*Corvus frugivorus* to *Ortyx virginiana*), July September, 1880.

The high praise accorded the earlier instalments is equally merited by those now under notice, Mr. Mearns's "List of the Birds of the Hudson Highlands" ranking easily among the best of our long list of contributions to local ornithology.... In respect to nomenclature, the list is abreast with the latest well-grounded innovations.—J. A. A., *Bull. Nutt. Ornith. Club*, Vol. VI., p. 172, July, 1881.

MINOT, H. D.—The Diary of a Bird. By H. D. Minot. Boston: A. Williams & Co., 1880, 8vo., pp. 38, cuts.

This entertaining and pleasantly written piece of bird-gossip is represented to be a translation of a "Diary" of a "Black-throated Green Warbler," and recounts, among other things the doings of "a grand mass meeting" of the birds to discuss "The Destruction and Extermination of Birds; how caused and how to be prevented," in which various members of the great bird convention relate their grievances....The object of this attractive little brochure is to awaken popular interest in the general subject of the better protection of our birds, not only against the professional market gunner, but from their wholesale destruction to meet the demands of the milliner.—J. A. A., *Bull. Nutt. Ornith. Club*, Vol. V., p. 112, April, 1880.

NEHRLING, H.—Ornithologische Beobachtungen aus Texas. I. Von H. Nehrling. Monatsschrift des Deutschen Vereins zum Schutze der Vogelwelt, V Jahrgang, No. 7, Juli, 1880, pp. 122-139.

These observations consist of a running commentary on the more common birds met with by Dr. Nehrling in March, April, and May, 1879, in Lee and Fayette Counties, Texas. It is apparently the first of a series of papers on the birds of Texas....with, incidentally, notes on the mammals, the plants, and the general character of the country.... --J. A. A., *Bull. Nutt. Ornith. Club*, Vol. VI., p. 109, April, 1881.

OBER, FREDERICK A.—Camps in the Caribbees: The Adventures of a Naturalist in the Lesser Antilles. By Frederick A. Ober. Boston: Lee and Shepard. New York: Charles T. Dillingham. 1880. 8vo., pp. xviii, 366, with 34 illus.

....The general text introduces a good deal of ornithological matter, which will be found of interest and value, and the appendix is entirely devoted to this subject. It gives Mr Lawrence's summary list of the species, 128 in number.... and also reproduces the original descriptions of all the new species discovered by the energetic and successful explorer.--E. C., *Bull. Nutt. Ornith. Club*, Vol. V., p. 179, July, 1880.

REICHENOW, ANTON, and SCHALOW, HERMANN.-- Compendium der neu beschriebenen Gattungen und Arten. Von Anton Reichenow und Hermann Schalow. *Journal für Ornithologie*, 1879, pp. 308-329, 420-437; 1880, pp. 97-102, 194-209, 314-324.

The authors of the "Compendium" are placing ornithologists under a debt of gratitude in promply bringing together the diagnoses of the new genera and species of current ornithological literature. The last instalment apparently covers the first half of the year 1880, and the families from *Cuculidæ* upward through the *Oscines*....--J. A. A., *Bull. Nutt. Ornith. Club*, Vol. VI., p. 111, April, 1881.

REICHENOW, ANTON, and SCHALOW, HERMANN.—Zoologischer Jahresbericht für 1879. Herausgegeben von der Zoologischen Station zu Neapel. Redigirt von Prof. J. Victor Carus (W. Englemann, Leipzig). 5. Aves. Bd. II., pp. 1108 1161. Referenten Dr. Ant. Reichenow und H. Schalow.

....The report appears to be very carefully and satisfactorily prepared, the annotations being sufficiently full and explicit. J. A. A., *Bull. Nutt. Ornith. Club*, Vol. VI , p. 111, April, 1881.

ROBERTS, THOMAS S.—The Convolutions of the Trachea in the Sandhill and Whooping Cranes. By Thomas S. Roberts, M.D. *American Naturalist*, Vol. XIV., February, 1880, pp. 108 114, figg.

... Mr. Roberts has given an admirable presentation of the tracheal characters of our two larger species of Cranes, illustrated with cuts .... J. A. A., *Bull. Nutt. Ornith. Club*, Vol. V., pp. 179, 180, July, 1880.

STEARNS, WINFRID A.—List of Birds of Fishkill on Hudson. N. Y. By Winfrid A. Stearns. 8vo., pp. 16. without date or publisher's impress. Published July 10, 1880.

This is a briefly annotated list of about 130 species, based on ten months' observations by the author in the vicinity of Fishkill....the list, though very incomplete, is doubtless trustworthy .. J. A. A., *Bull. Nutt. Ornith. Club*, Vol. V , p. 233, October, 1880.

STEERE, J. B.—A List of the Mammals and Birds of Ann Arbor and Vicinity. By Professor J. B. Steere. 8vo., pp. 8, 1880.

This briefly annotated list of 111 species (of birds) is good as far as it goes "....with the exception of a few, given upon the authority of labeled specimens in the Museum, it is the result of about three years' collecting and observation in this vicinity." J. A. A., *Bull. Nutt. Ornith. Club*, Vol. VI., p. 46, January, 1881.

1881.

BAILEY, H. B.—"Forest and Stream" Bird Notes. An index and summary of all the ornithological matter contained in "Forest and Stream," Vols. 1-XII. Compiled by H. B. Bailey. New York: F. & S. Pub. Co., 39 Park Row, 1881. 8vo., paper, pp. iv., 195.

. His work is more than a mere alphabetical list of names, followed by reference figures ; for it includes .. a summary of each article indexed ...The Index also includes author's names, and among these the authorship of many pseudonyms and initial signatures are for the first time properly exposed. The summation of the bird-matters seems to be quite complete and is certainly extensive... - E. C., *Bull. Nutt. Ornith. Club*, Vol. VII., pp. 175, 176, July, 1882.

FREKE, PERCY EVANS.—On Birds observed in Amelia County, Virginia. By Percy E. Freke. *Scientific Proc. Royal Dublin Society*, Vol. III. Part III. [Read February 21, 1881.]

.. Mr. Freke has done good service in publishing the results of six years' observations in Amelia County, at a point about thirty miles south of Richmond. His list, which is freely annotated, includes 112 species ...The author has evidently fallen into some confusion regarding the spotted-breasted Thrushes of the genus *Turdus* ...will be read with interest, not only as an exponent of the ornithology of a previously unworked section, but also as embodying a foreigner's pleasantly told impressions of many of our familiar birds W. B., *Bull. Nutt. Ornith Club*, Vol. VII., p. 48, January, 1882.

FREKE, PERCY EVANS.—North American Birds crossing the Atlantic. By Percy Evans Freke. 8vo., pp. 11. *Scientific Proc. Royal Dublin Society*, Vol. III., 1881.

This paper is based on the author's "Comparative Catalogue of Birds found in Europe and North America"... of which it may be regarded as in part a summary, as also a most valuable résumé of the general subject of North American birds occurring in Europe. The number of species is 69; the total number of occurrences, 494... —J. A. A., *Bull. Nutt. Ornith. Club*, Vol. VIII., pp. 114, 115, April, 1883.

FREKE, PERCY EVANS.—On European Birds observed in North America. By Percy E. Freke. *Zoölogist*, September, 1881.

The total number of species included in this list is 56, of which 9 are regarded as artificially introduced....The list seems to have been most carefully worked out, and may deservedly stand as a companion piece to Mr. J. J. Dalgleish's "List of Occurrences of North American Birds in Europe," published in Volume V. of this Bulletin ....—J. A. A., *Bull. Nutt. Ornith. Club*, Vol. VIII., p 115, April, 1883.

GARROD, ALFRED HENRY, and FORBES, W. A.—In Memoriam. The Collected Scientific Papers of the late Alfred Henry Garrod, M.D., F.R.S., etc. Edited, with a biographical memoir of the author, by W. A. Forbes, B.A., etc. London: R. H. Porter, 6 Tenterden Street. 1881. 1 vol., 8vo., pp. xxvi., 538, pll. 33, frontispiece (portrait) and many cuts in text.

...Of the anatomical papers in the present volume, some 73 in number, more than half relate to birds, describing conditions of the osseous, muscular, respiratory, vascular, digestive, and nervous systems ....and discussing in candid and scientific spirit .. the bearing of the anatomical points upon classification. Of the accuracy and high rate of reliability of these papers there can be no question among them is an entirely new classification of birds, based primarily upon the ambiens [muscle]....—E. C., *Bull. Nutt. Ornith. Club*, Vol. VII., pp. 43, 44, January, 1882.

GODMAN, F. DUCANE, and SALVIN, OSBERT.—Biologia Centrali-Americana; or, Contributions to the knowledge of the Fauna and Flora of Mexico and Central America. Edited by F. Ducane Godman and Osbert Salvin. Zoölogy, Parts I-X. Aves, by O. Salvin and F. D. Godman, pp. 1–152, pll. i-x. 4to. London: Published for the Editors by R. H. Porter, 10 Chandos Street,

Cavendish Square, W., and Dulau & Co., Soho Square. September, 1879 April, 1881.

....As the title indicates, the work treats of the fauna and flora of Mexico and Central America ...The ornithological portion is by the editors ..Of each species a short Latin description is given, and all the more important references to the literature are duly cited. ..The ten plates thus far published contain figures of 25 hitherto unfigured species ...The importance and usefulness of the present work cannot ....be easily overestimated....The execution of the "Biologia" as regards typography and illustrations.. is excellent ... J. A. A., *Bull. Nutt Ornith. Club*, Vol. VII., pp 174 176, July, 1881.

HARVIE-BROWN, JOHN A., CORDEAUX, JOHN, and KERMODE, PHILIP.—Report on the Migration of Birds in the Spring and Autumn of 1880. By John A. Harvie-Brown, F.L.S., F.Z.S., John Cordeaux, and Philip Kermode. London: W. S. Sonnenschein & Allen, 15, Paternoster Square. 1881. 8vo., pp. 120.

....we now....call attention to several late reports and papers on the same subject [migration of birds]. The report for 1880 forms a pamphlet of 120 octavo pages ...printed schedules and letters of instruction were sent to 39 stations ...on the east coast of Scotland . to 44 on the east coast of England ; to 38 on the west coast of Scotland ....and to 39 on the west coast of England, or to 160 stations in all. from 106 of which reports were received... The report for 1881 is of similar scope and character....—J. A. A., *Bull. Nutt. Ornith. Club*, Vol. VIII., pp. 228, 229, October, 1883.

HARVIE-BROWN, JOHN A.—Paper on the Migration of Birds upon our British Coasts, read before the Stirling Field Club, on Tuesday, 13th December, 1881, by J. A. Harvie-Brown, F.R.S.E., F.Z.S., etc. Stirling: Printed at the Journal and Advertiser Office. 1881. 12mo., pp. 12.

HATCH, P. L.—A List of the Birds of Minnesota. By Dr. P. L. Hatch. *Ninth Ann. Rep. Geol. and Nat. Hist. Surv. Minn., for* 1880. 1881, pp. 361-372.

....a list of 281 species, briefly annotated ... E. C., *Bull. Nutt. Ornith. Club*, Vol. VII., p. 47, January, 1882.

HOLTERHOFF, G., JR.—A Collector's Notes on the Breeding of a few Western Birds. By E. [*i. e.*, G.] Holterhoff, Jr. *American Naturalist*, March, 1881, pp. 208-219.

...The observations here recorded were made in Southern California in the spring of 1880 and have reference to some 40 species.... J. A. A., *Bull. Nutt. Ornith. Club*, Vol. VI., p. 173, July, 1881.

HOFFMAN, W. J.—Annotated List of the Birds of Nevada. By W. J. Hoffman, M.D., *Bull. U. S. Geol. and Geogr. Surv. of Terr.*, Vol. VI., No. 2, Sept. 19, 1881, pp. 203-256, and Map.

....The list is based partly upon the writer's personal experience in the field during the season of 1871, but mainly upon .. previously published reports ...It hence partakes largely of the nature of a compilation, although the author's original notes are by no means few or uninteresting....The paper ... closes with a bibliographical list of the chief publications relating to the region considered, and an excellent map of the State....Dr. Hoffman's paper ...should find a place in the hands of every working ornithologist.—W. B., *Bull. Nutt. Ornith. Club*, Vol. VII., p. 51, January, 1882.

KRUKENBERG, C. FR. W.—Die Farbstoffe der Federn, in dessen vergleichend-physiologische Studien. Von Dr. C. Fr. W. Krukenberg. I Reihe, V Abth., 1881, pp. 72-92. Plate iii.

This paper, the first of a series, seems to be the product of more careful work than previous publications on the subject [coloring matter of feathers]....—J. Amory Jeffries, *Bull. Nutt. Ornith. Club*, Vol. VII., pp. 114, 115, April, 1882.

LANGDON, F. W.—Field Notes on Louisiana Birds. By Dr. F. W. Langdon. *Journ. Cincinnati Soc. Nat. Hist.*, July, 1881, pp. 145 155.

...."a record of....the month ending April 17, 1881 at 'Cinclaire' ....in the parish of West Baton Rouge"....the paper will be welcomed as an acceptable contribution to our knowledge of a region which has been nearly a *terra incognita* to ornithologists since the days of Audubon. —W. B., *Bull. Nutt. Ornith. Club*, Vol. VII., pp. 40, 49, January, 1882.

LANGDON, F. W.—Zoölogical Miscellany, edited by Dr. F. W. Langdon. *Jour. Cincinnati Soc. Nat. Hist.*, Vol. IV., Dec., 1881, pp. 336–346.

...."facts....respecting the structure, the life history, or the geographical distribution of the various species of animals constituting the Ohio Valley Fauna." The part before us includes sections on mammalogy, ornithology, herpetology, ichthyology, conchology, and entomology....the editor contributes a short but useful paper on the "Introduction of European Birds" ... W. B., *Bull. Nutt. Ornith. Club*, Vol. VII., pp. 50, 51, January, 1882.

LAWRENCE, GEORGE N.—Description of a New Subspecies of Loxigilla from the Island of St. Christopher, West Indies. By George N. Lawrence. *Proc. U. S. Nat. Mus.*, Vol. IV., 1882, pp. 204, 205.

....Mr. Lawrence describes a new subspecies of Loxigilla (*L. portoricensis* var *grandis*)....—J. A. A., *Bull. Nutt. Ornith Club*, Vol. VIII., p. 114, April, 1883.

MACOUN, JOHN.—Extract from a Report of Exploration by Professor John Macoun, M.A., F.L.S. *Report of Department of Interior* (Ottawa, 1881 ?) 8vo., pp. 48.

....chiefly (pp. 8 40) of Professor Macoun's own report of his explorations during the summer of 1880....north of our territories of Dakota and Montana....the present paper possesses decided value, as the observer appeared to have paid special attention to the distribution of birds in the wide area traversed. After a résumé of the leading ornithological features of the region is presented an annotated list of the species secured, 109 in number ... We feel at liberty to call attention to some manuscript alterations made by the author in our copy. For *Coturniculus passerinus* read *Zonotrichia albicollis*; for *Myiarchus crinitus*, read *Tyrannus verticalis*; for *Archibuteo lagopus*, read *A. ferrugineus* ... for *Tringa canutus* read *T. bairdi*; for *Podilymbus podiceps*, read *Podiceps californicus*.. .—E. C., *Bull. Nutt. Ornith. Club*, Vol. VII., p. 113, April, 1882.

RATHBUN, FRANK R.—Bright Feathers or some North American Birds of Beauty. By Frank R. Rathbun. Illustrated with Drawings from Nature, and carefully colored by hand. Auburn, N. Y. Published by the Author, 1880. 4to. Part I., pp. i-viii, 9-24, colored Plate and colored Vignette.

....is an attractive piece of book making; the drawing of the plate is passable, and the coloring is not more highly exaggerated than in many plates by authors of reputation for accuracy. The text more clearly betrays the hand of inexperience .. The author is evidently not wanting in knowledge of his subject; the faults of style he will doubtless be able to overcome as the work proceeds. . J. A. A., *Bull. Nutt. Ornith. Club*, Vol. V., p. 234, October, 1880.

——— Part II.

Part II. of this work,....is devoted to the Rose-breasted Grosbeak (*Goniaphea ludoviciana.*) The colored plate illustrates the adult male and female, but the sixteen quarto pages (pp. 25-40) of text leave the history of the species still unfinished. In noticing Part I ... we were compelled to speak unfavorably of the literary execution of the work, and regret that the present issue will not permit of a more favorable notice ....- J. A. A., *Bull. Nutt. Ornith. Club*, Vol. VI., pp. 172, 173, July, 1881.

RIDGWAY, ROBERT.—Revisions of Nomenclature of certain North American Birds. By Robert Ridgway. *Proc. U. S. Nat. Mus.*, Vol. III., 1881, pp. 1-16. Author's separates issued March 27, 1880.

....Mr. Ridgway takes as a starting-point Dr. Coues's "Check List" published in 1873, and formally notices many of the changes from the nomenclature there adopted ...and proposes many additional ones, the whole number here receiving attention amounting to upward of eighty....—J. A. A., *Bull. Nutt. Ornith. Club*, Vol. V., pp. 177, 193, July, 1880.

RIDGWAY, ROBERT.—Nomenclature of North American Birds chiefly contained in the United States National Museum. By Robert Ridgway. *Bull. U. S. Nat. Mus.*, No. 21. Published under the direction of the Smithsonian Institution. Washington: Government Printing Office, 1881. 8vo., pp. 1-94.

....its publication marks an epoch in North American ornithology....The actual number of names ...in the present catalogue (1881),

"924"....the system is trinominal....The work....evinces the exercise of the utmost care in its preparation. J. A. A., *Bull. Nutt. Ornith. Club*, Vol. VI., pp. 164 171, July, 1881.

RIDGWAY, ROBERT.—A Revised Catalogue of the Birds ascertained to occur in Illinois. By Robert Ridgway. *Illinois State Laboratory of Natural History. Bulletin No.* 4. Bloomington, Ill., May, 1881. 8vo., pp. 161–208.

....based primarily upon the same author's "Catalogue of the Birds ascertained to occur in Illinois," published....in 1874, but adds 31 species....341 now enumerated, besides 11 additional varieties.... The species known to breed (213 in number) are distinguished by an asterisk....The nomenclature is that of Mr. Ridgway's recently published "Catalogue of North American Birds" .. Illinois takes the lead among the States in respect to number of species of birds ...—J. A. A., *Bull. Nutt. Ornith. Club*, Vol. VI , pp. 171, 172, July, 1881.

ROBERTS, THOMAS S.—The Winter Birds of Minnesota. By Thomas S. Roberts. *Ninth Ann. Rep. Geol. and Nat. Hist. Surv. Minn., for* 1880, 1881, pp. 373–383.

....treats....of 52 species known to occur in the State in winter ....the information given conveying a good idea of the bird-fauna at that season of the year....—E. C., *Bull. Nutt. Ornith. Club*, Vol. VII., p. 47, January, 1882.

SEEBOHM, HENRY.—Catalogue of the Birds in the British Museum. Vol. V. Catalogue of the Passeriformes, or Perching Birds in the British Museum. Cichlomorphæ: Part II., containing the family Turdidæ (Warblers and Thrushes). By Henry Seebohm. London, 1881. 8vo., pp. xvi, 426, pll. xviii.

....this group is defined in Mr. Sharpe's scheme of classification, with limits rather different from those usually assigned to it ...we admire most heartily his [Mr. Seebohm's] thorough treatment of the subject in hand and the philosophic spirit in which he has approached his task....—J. A. A., *Bull. Nutt. Ornith. Club*, Vol. VIII., pp. 99–104, April, 1883.

SHARPE, R. BOWDLER.—Catalogue of the Birds in the British Museum. Vol. VI. Catalogue of the Passeriformes, or Perching Birds, in the collection of the British Museum. Cichlomorphæ: Part III., containing the first portion of the family Timeliidæ (Babbling Thrushes). By R. Bowdler Sharpe. London, 1881. 8vo., pp. xiii, 420, pll. xviii.

....In respect to the classification followed in these volumes, Mr. Sharpe states that it is based on that of the late Professor Sundevall. ....—J. A. A., *Bull. Nutt. Ornith. Club*, Vol. VIII., pp. 104, 105, April, 1883.

SHUFELDT, R. W.—Osteology of Speotyto cunicularia var. hypogæa. By R. W. Shufeldt, [First Lieutenant and] Assistant Surgeon,

U. S. Army. *Bull. U. S. Geol. and Geogr. Surv. of Terr.*, Vol. VI., No. 1, February 11, 1881, pp. 87–117, pll. i–iii.

SHUFELDT, R. W.—Osteology of Eremophila alpestris. By R. W. Shufeldt, [First Lieutenant and] Assistant Surgeon, U. S. Army. *Bull. U. S. Geol. and Geogr. Surv. of Terr.*, Vol. VI., No. 1, February 11, 1881, pp. 119–147, pl. iv.

As memoirs of descriptive osteology these papers merit high praise, and may well be welcomed as valuable contributions in a little worked field.—J. A. A., *Bull. Nutt. Ornith. Club*, Vol. VI., pp. 109, 110, April, 1881.

SHUFELDT, R. W.—Osteology of the North American Tetraonidæ. By Dr. R. W. Shufeldt, U. S. A. *Bull. U. S. Geol. and Geogr. Surv. of Terr.*, Vol. VI., No. 2, pp. 309–350, pll. v–xiii.

...so far as we know, the most complete of any [paper] on American birds of one group ...—J. Amory Jeffries, *Bull. Nutt. Ornith. Club*, Vol. VII., pp. 44, 45, January, 1882.

SHUFELDT, R. W.—Osteology of Lanius ludovicianus excubitoroides By Dr. R. W. Shufeldt, U. S. A. *Bull. U. S. Geol. and Geogr. Surv. of Terr.*, Vol. VI., No. 2, pp. 351–359, pl. xiv.

The description ...is short, concise, and may be summed up in the statement that the skeleton of this bird is strictly Passerine.—J. Amory Jeffries, *Bull. Nutt. Ornith. Club*, Vol. VII., p. 45, January, 1882.

SHUFELDT, R. W.—The Claw on the Index Digit of the Cathartidæ. By R. W. Shufeldt, M.D. *American Naturalist*, November, 1881, pp. 906–908.

....this paper contains such important errors, both in regard to structure of birds and the literature of the subject that some rectification seems necessary. Dr. Shufeldt describes the claw at the end of the first finger of *Catharista atrata* as a new discovery, considering that claws outside of the Ostrich groups have not hitherto been described, and also states that it is a point of distinction between the Old and New World Vultures. ...the claw on the first finger is anything but an unknown object ...That the claw is absent in the Old World Vultures is also an error if we may trust the high authority of Nitzsch... as a rule the claws are much more conspicuous in young than in adult birds.- J. Amory Jeffries, *Bull. Nutt. Ornith. Club*, Vol. VII., pp. 46, 47, January, 1882.

STEARNS, WINFRID A., and COUES, ELLIOTT.—New England Bird Life, being a Manual of New England Ornithology, revised and edited from the manuscript of Winfrid A. Stearns, Member of the Nuttall Ornithological Club, etc., by Dr. Elliott Coues, U. S. A., Member of the Academy, etc. Part I.-Oscines. Boston: Lee and Shepard, Publishers. New York: Charles T. Dillingham. 1881. 8vo., pp. 324, numerous woodcuts.

....we at length have a work on New England Birds of which no ornithologist need feel ashamed... The main body of the work comprises two hundred and seventy pages and treats....the whole order *Oscines*....The claims of each species to be considered a member of the New England Fauna are critically examined....the design being to give a thoroughly reliable list of the Birds, with an account of the leading facts in the life-history of each species. The plan of the work includes brief descriptions of the birds themselves, enabling one to identify any specimen ...To say that the book is exceedingly well-written would be doing it scant justice. Dr. Coues's brilliant talents in this respect are already well known, but we have perhaps never had so striking a proof of them as is afforded by the present volume......Mr. Stearns may be congratulated on his wise choice of an editor. -W. B., *Bull. Nutt. Ornith. Club*, Vol. VI., pp. 236-240, October, 1881.

1882.

BICKNELL, EUGENE PINTARD.—A Review of the Summer Birds of a part of The Catskill Mountains, with prefatory remarks on the faunal and floral features of the region. By Eugene Pintard Bicknell. *Transactions of the Linnæan Society of New York.* Vol. I., pp. 113-168, December, 1882.

....is based on observations made "during brief explorations of the more southern Catskills in three successive years, from June 6 15, 1880; 12-18, 1881; 24-27, 1882....Twenty-five of the total fifty-six pages are devoted to prefatory remarks....Mr. Bicknell evidently has a penchant for the analysis and comparison of faunæ, and his remarks in the present connection are decidedly interesting .. The list proper includes eighty-nine species and varieties. It is very fully annotated. ....—W. B., *Bull. Nutt. Ornith. Club*, Vol. VIII., p. 53, January, 1883.

BLASIUS, RUDOLPH.—V. Jahresbericht (1880) des Ausschlusses für Beobachtungs-stationen der Vögel Deutschlands. Journal für Ornithologie, XXX Jahrg., Heft I, Jan., 1882, pp. 18-110.

The fifth annual report of the German observers for the year 1880 ...is presented in the form of an annotated list of 280 species, compiled from the reports of the various observers ...The notes relate to nesting of many of the species, as well as to their migrations .. There are ...reports from no less than 36 stations, and the résumé of the observations taken forms a paper of great interest and value.—J. A. A., *Bull. Nutt. Ornith. Club*, Vol. VIII., pp. 229, 230, October, 1883.

BROWN, NATHAN CLIFFORD.—A Catalogue of the Birds known to occur in the vicinity of Portland, Me. [etc.] By Nathan Clifford Brown. *Proc. Portland Soc. Nat. Hist.*, Dec. 4, 1882.

This excellent local list....is stated to be prepared from notes systematically taken during the past twelve years, and to contain the names of scarcely any species which have not passed under the author's personal observation. Its reliability is therefore evident. The number of species given is 250 ...The annotations, though not extensive, are to the point and seem judiciously adapted to convey a fair idea of the part each species plays in the composition of the Avifauna. .. —E. C., *Bull. Nutt. Ornith. Club*, Vol. VIII., pp. 112, 113, April, 1883.

CHAMBERLAIN, MONTAGUE.—A Catalogue of the Birds of New Brunswick, with brief notes relating to their migrations, breeding, relative abundance, etc. By Montague Chamberlain. *Bulletin of the Natural History Society of New Brunswick.* No. 1, pp. 23-68. Published by the Society. Saint John, N. B., 1882.

..This paper ...comprises some forty-three pages, which are divided into two sections; "Section A" being restricted to species which occur in St. John and King's Counties; while "Section B" embraces "species which have not been observed in Saint John or King's Counties but which occur in other parts of the Province." The former division treats of a region to which the author has evidently paid special attention, and the text, being mainly based on his personal observations or investigations, includes many interesting and several important notes and records... Section B is almost wholly compiled .. Mr. Chamberlain's work, so far as it has gone, has evidently been done carefully and well....in many respects it lacks the completeness that is desirable in a paper of its kind ...—W. B., *Bull. Nutt. Ornith. Club*, Vol. VII., pp. 176, 177, July, 1882.

COLLINS, J. W.—Notes on the Habits and Methods of Capture of various species of Sea Birds that occur on the Fishing Banks off the Eastern Coast of North America, and which are used as bait for catching Codfish by New England Fishermen. By Capt. J. W. Collins. *Ann. Rep. of the Comm. of Fish and Fisheries for* 1882, pp. 311-338, pl. i.

... particularly welcome, not only for the information they convey on these points [sea-birds captured and used as bait], but also respecting the relative abundance of the sea-birds met with on the fishing banks, their habits, seasons of occurrence, and migrations.... The species captured in largest numbers is the Greater Shearwater (*Puffinus major*)....—J. A. A., *The Auk*, Vol. I., pp. 380, 381, October, 1884.

COUES, ELLIOTT.—The Coues Check List of North American Birds, revised to date and entirely rewritten under direction of the author, with a Dictionary of the Etymology, Orthography and Orthoëpy of the scientific names, the Concordance of previous lists, and a Catalogue of his Ornithological Publications. Boston: Estes and Lauriat. 1882. 1 vol. Royal 8vo., pp. 165.

... it is much more than a catalogue of North American birds.... the erudition and scholarly research involved in this undertaking must be apparent to the most casual reader. The practical value of the work is equally plain....The total number of species and varieties enumerated is eight hundred and eighty-eight ... W. B., *Bull. Nutt. Ornith. Club*, Vol. VII., pp. 111, 112, April, 1882.

....The purpose of the present 'Check List' is, First to present a complete list of the birds now known to inhabit North America, north of Mexico and including Greenland .. Secondly to take each word.... explain its derivation, significance, and application, spell it correctly and indicate its pronunciation .. Concerning the whole work we can say nothing stronger than that it is in every way worthy of its brilliant and distinguished author, who has evidently made it one of his most mature and carefully studied efforts....it fills a field of usefulness peculiarly its own.....—W. B., *Bull. Nutt. Ornith. Club*, Vol. VII., pp. 212–216, October, 1882.

Dubois, Alphonse.—De la Variabilité des Oiseaux du genre Loxia. Par M. Alph. Dubois, Conservateur au Musée royal d'histoire naturelle de Belgique. *Extrait du Bulletin du Musée royal d'histoire naturelle de Belgique.* Tome I. Oct., 1882.

....These varieties, races, or subspecies, he holds to be the result of the action of climate, food, or other "fortuitous causes" upon size and coloration, and states that his morphological studies have demonstrated that species are variable in proportion to the extent of their area of dispersion.....—J. A. A., *Bull. Nutt. Ornith. Club*, Vol. VIII., p. 170, July, 1883.

Dutcher, William.—Is Not the Fish Crow (Corvus ossifragus Wilson) a winter as well as a summer resident at the northern limit of its range? By William Dutcher. *Transactions of the Linnæan Society of New York.* Vol. I., pp. 107–111, December, 1882.

....is short, occupying less than three pages....The evidence cited is apparently conclusive....—W. B., *Bull. Nutt. Ornith. Club*, Vol. VIII., p. 54, January, 1883.

Forbes, S. A.—The Regulative Action of Birds upon Insect Oscillations. By S. A. Forbes. *Bull. No. 6, Illinois State Laboratory of Nat. Hist.*, Dec., 1882, pp. 1–31.

Our best authority upon the insect food of birds has continued his observations upon the subject ...The paper is very carefully worked up to show how effectively birds may restore a disturbed balance of life ....We trust Professor Forbes will not desist from his good work. Such exact data as these are just what is required for the solution of the general problem which is offered by the relations of the bird-world to agriculture.—E. C., *Bull. Nutt. Ornith. Club*, Vol. VIII., pp. 105–107, April, 1883.

Gentry, Thomas G.—Nests and Eggs of the Birds of the United States [Pennsylvania]. 4to. 50 colored Plates. 1882.

Part I. of this new enterprise....has reached us .. The text of this number is meritorious, and the plates are not....—E. C., *Bull. Nutt. Ornith. Club*, Vol. V., p. 179, July, 1880.

Harvie-Brown, John A., Cordeaux, John, and Newton, Alfred.—Report of the Committee, consisting of Mr. J. A. Harvie-Brown, Mr. John Cordeaux, and Professor Newton, appointed at Swansea "for the purpose of obtaining (with the consent of the Master and Brethren of the Trinity House, and of the Commissioners of Northern Lights) observations on the Migration of Birds at Lighthouses and Lightships, and of reporting on the same, at York, in 1881." London: Printed by Spottiswoode and Co., New-Street Square and Parliament Street. [1882.] 8vo., pp. 8.

Harvie-Brown, John A., [etc.]—Report on the Migration of Birds in the Autumn of 1881. By John A. Harvie-Brown, Mr. John Cordeaux, Mr. Philip M. C. Kermode, Mr. R. M. Barrington, and Mr. A. G. More. London: Printed by West, Newman & Co., 54, Hatton Garden. 1882, 8vo., pp. 101.

Hoffman, W. J.—List of Birds observed at Ft. Berthold, D. T., during the month of September, 1881. By W. J. Hoffman, M.D. *Proc. Boston Soc. Nat. Hist.*, Feb. 1, 1882.

....the result of some observations made during September, 1881 ....Fifty-seven species were identified . The annotations are usually very brief ...A novel feature of the list is that of the Indian names which are given for many of the common birds . W. B., *Bull. Nutt. Ornith. Club*, Vol. VIII., pp. 54, 55, January, 1883.

Ingersoll, Ernest.—Birds'-Nesting: A Handbook of Instruction in Gathering and Preserving the Nests and Eggs of Birds for the Purposes of Study. By Ernest Ingersoll. Salem, 1882.

This little book is intended for a guide to the beginner.. The book may be summarized as a readable account of the various modes of collecting birds' eggs and nests A long account of the various paraphernalia for blowing and marking eggs is given... A list of unknown nests contains faults of admission, though these are not numerous ....—J. A. J., *Bull. Nutt. Ornith. Club*, Vol. VII., pp. 179, 180, July, 1882.

Knowlton, F. H.—A Revised List of the Birds of Brandon, Vt., and vicinity. By F. H. Knowlton. The Brandon Union (newspaper), February 10, 1882.

This is a briefly annotated list of 149 species.... The chief interest of the list lies in its bearing upon the extent of the Alleghanian fauna in the Champlain valley....Mr. Knowlton has recorded Wilson's Plover ....instead of Wilson's Snipe.—C. F. B., *Bull. Nutt. Ornith. Club*, Vol. VII., pp. 113, 114, April, 1882.

Krukenberg, C. Fr. W.—Die Farbstoffe der Federn in Dessen vergleichend-physiologische Studien. Von Dr. C. Fr. W. Krukenberg. II Reihe, I Abth., 1882, pp. 151, 171.

....the author describes the yellow pigment, Coriosulfurin, found in the tarsus of the birds of prey ....—J. A. J., *Bull. Nutt. Ornith. Club*, Vol. VII., pp. 177, 178, July, 1882.

Lawrence, George N.—Description of a New Species of Swift of the genus Chætura, with Notes on two other little-known Birds. By George N. Lawrence. *Ann. New York Acad. Sci.*, Vol. II., No. 8, pp. 247, 248. March, 1882.

Lawrence, George N.—Descriptions of New Species of Birds from Yucatan, of the Families Columbidæ and Formicariidæ. By George N. Lawrence. *Ann. New York Acad. Sci.*, Vol. II., No. 9, pp. 287, 288. May, 1882.

LAWRENCE, GEORGE N.—Description of a New Species of Bird of the Family Cypselidæ. By George N. Lawrence. *Ann. New York Acad. Sci.*, Vol. II., No. 11, pp. 355, 356. December, 1882.

LINDEN, CHARLES.—On the Domestication of some of our Wild Ducks. By Charles Linden. *Bull. Buffalo Soc. Nat. Sciences*, Vol. IV., No. 2, pp. 33–39, 1882.

After brief reference to the various species of wild Ducks that formerly frequented Lake Chautauqua, Western New York, which have now mostly become rare, Mr. Linden summarizes the results of systematic efforts continued for nearly thirty years by Mr. George Irwin at the above-named locality to domesticate several of the species. These were the Mallard, Dusky Duck, Wood Duck, Blue-winged Teal, and American Swan. All of these bred freely and reared their young in confinement....—J. A. A., *Bull. Nutt. Ornith. Club*, Vol. VIII., p. 233, October, 1883.

MERRIAM, CLINTON HART.—The Vertebrates of the Adirondack Region, Northeastern New York. By Clinton Hart Merriam, M.D. [First Instalment.] *Transactions of the Linnæan Society of New York.* Vol. I., pp. 5–106, December, 1882.

....The present instalment of Dr. Merriam's paper does not extend to birds....its introductory portion has a direct bearing on everything to follow....As a contribution to our knowledge of the habits, food, times and manner of breeding, etc., of many of the northern mammals this paper is an important one ....—W. B., *Bull. Nutt. Ornith. Club*, Vol. VIII., pp. 50–53, January, 1883.

MORDEN, J. A., and SAUNDERS, W. E.—List of the Birds of Western Ontario. By J. A. Morden and W. E. Saunders. *Canadian Sportsman and Naturalist*, Vol. II., Nos. 11 and 12, pp. 183–187, 192–194. November and December, 1882.

....a briefly annotated list ...numbering 236 species ...a valuable addition to our knowledge of the distribution of Canadian birds....—J. A. A., *The Auk*, Vol. I., p. 85, January, 1884.

REICHENOW, ANTON.—Conspectus Psittacorum. Systematische Uebersichte aller bekannten Papageienarten. Von Dr. Ant. Reichenow. 8vo., Berlin, 1882, pp. 234. (Sonderabdruck aus Journal für Ornithologie, XXIX Jahrg., 1881, pp. 1–49, 113–177, 225–289, 337–398.)

The order *Psittaci* is divided into 9 families and 45 genera (including 27 subgenera); 444 species and subspecies are recognized .. English and French, as well as German, vernacular names are given . It originally appeared in parts in the "Journal für Ornithologie" for 1881.—J. A. A., *Bull. Nutt. Ornith. Club*, Vol. VIII., p. 169, July, 1883.

REICHENOW, ANTON.—Die Vögel der Zoologischen Gärten. Leitfaden zum Studium der Ornithologie mit besonderer Berücksichtigung der in Gefangenschaft gehaltenen Vögel. Ein Handbuch für

Vogelwirthe. Von Dr. Ant. Reichenow. In zwei Theilen. [Theil I.] Leipzig, 1882, 8vo., pp. xxx., 278.

Dr. Reichenow's handbook for bird-keepers is designed to furnish ....the means of readily identifying such species as are kept in zoölogical gardens, parks, and aviaries, and seems to be well adapted to that end. The first part . treats of 695 species .. Concise diagnoses are given .. and English and French, as well as German, vernacular names are supplied for the species. As a popular hand book for German readers .. the work seems worthy of generous commendation. J. A. A., *Bull. Nutt. Ornith. Club*, Vol VIII., p. 232, October, 1883.

REICHENOW, ANTON.- Die Entenvögel der Zoologischen Gärten. Von Ant. Reichenow. Ornithologisches Centralblatt. VII Jahrg., Nos. 1-6. Jan.-May, 1882, pp. 1-5, 17-23, 35-40.

....enumerates the species of *Lamellirostres*... giving brief diagnoses of the species kept in zoölogical gardens .. J. A. A., *Bull. Nutt. Ornith. Club*, Vol. VIII., p. 232, October, 1883.

REICHENOW, ANTON, and SCHALOW, HERMAN. Compendium der neu beschriebenen Gattungen und Arten. Von Anton Reichenow und Herman Schalow. Journal für Ornithologie, XXIX Jahrg., 1881, pp. 70-102, 417-423; XXX Jahrg., 1882, pp. 111-120, 213-228.

This convenient summary....is still continued....it gives transcripts of the original diagnoses, when such are given, and in other cases mentions the types of the genera and the alleged characteristics of the species.—J. A. A., *Bull. Nutt. Ornith. Club*, Vol. VIII., p. 169, July, 1883.

RIDGWAY, ROBERT.—On a Duck new to the North American Fauna. By Robert Ridgway. *Proc. U. S. Nat. Mus.*, Vol. IV., 1882, pp. 22-24. Author's separates issued April 13, 1881.

....Mr. Ridgway records an immature male Rufous-crested Duck (*Fuligula rufina*, Steph.) supposed to have been shot on Long Island Sound.. In making the record Mr. Ridgway takes occasion to describe the species in its various phases of plumage, and adds a few critical remarks on the generic synonomy of the group to which it belongs.—J. A. A., *Bull Nutt. Ornith. Club*, Vol. VI., p. 173, July, 1881.

RIDGWAY, ROBERT.—On Amazilia yucatanensis (Cabot) and A. cerviniventris, Gould. By Robert Ridgway. *Proc. U. S. Nat. Mus.*, Vol. IV., 1882, pp. 25, 26. Author's separates issued April 13, 1881.

....Comparative diagnoses are given of the two species, with some remarks respecting their distribution. J. A. A., *Bull. Nutt. Ornith. Club*, Vol. VI., pp. 173, 174, July, 1881.

RIDGWAY, ROBERT.—A Review of the genus Centurus, Swainson. By Robert Ridgway. *Proc. U. S. Nat. Mus.*, Vol. IV., 1882, pp. 93-119. Author's separates issued June 2, 1881.

This revision is based on an examination of 227 specimens, representing 12 of the 14 forms considered as sufficiently distinct for recognition....Each form recognized is described in detail, and the whole subject is treated with Mr. Ridgway's usual care and completeness. - J. A. A., *Bull. Nutt. Ornith. Club*, Vol. VIII., p. 114, April, 1883.

RIDGWAY, ROBERT.—List of Species of Middle and South American Birds not contained in the United States National Museum. By Robert Ridgway. *Proc. U. S. Nat. Mus.*, Vol. IV., 1882, pp. 165 203. Author's separates issued Aug. 11 and Nov. 18, 1881.

....The species wholly unrepresented are very few ...—J. A. A., *Bull. Nutt. Ornith. Club*, Vol. VIII., p. 170, July, 1883.

RIDGWAY, ROBERT.—List of Special Desiderata among North American Birds. By Robert Ridgway. *Proc. U. S. Nat. Mus.*, Vol. IV., 1882, pp. 207-223. Author's separates issued Nov. 18, 1881.

RIDGWAY, ROBERT.—Catalogue of Old World Birds in the United States National Museum. By Robert Ridgway. *Proc. U. S. Nat. Mus.*, Vol IV., 1882, pp. 317-333. Author's separates issued March 8, 1882.

....The numeration and classification adopted is that of Gray's well-known "Hand-list."—J. A. A., *Bull. Nutt. Ornith. Club*, Vol. VIII., p. 231, October, 1883.

RIDGWAY, ROBERT.—Notes on some Costa Rican Birds. By Robert Ridgway. *Proc. U. S. Nat. Mus.*, Vol. IV., 1882, pp. 333-337. Author's separates issued March 10, 1882.

RIDGWAY, ROBERT.—Description of a new Flycatcher and a supposed new Petrel from the Sandwich Islands. By Robert Ridgway. *Proc. U. S. Nat. Mus.*, Vol. IV., 1882, pp. 337, 338. Author's separates issued March 29, 1882.

RIDGWAY, ROBERT.—Description of a new Owl from Porto Rico. By Robert Ridgway. *Proc. U. S. Nat. Mus.*, Vol. IV., 1882, pp. 366-371. Author's separates issued April 6, 1882.

RIDGWAY, ROBERT.—Description of two new Thrushes from the United States. By Robert Ridgway. *Proc. U. S. Nat. Mus.*, Vol. IV., 1882, pp. 374-379. Author's separates issued April 6, 1882.

RIDGWAY, ROBERT.—On two Recent Additions to the North American Bird Fauna, by L. Belding. By Robert Ridgway. *Proc. U. S. Nat. Mus.*, Vol. IV., 1882, pp. 414, 415. Author's separates issued April 24, 1882.

In numerous papers published in the "Proceedings" of the National Museum for 1881 and 1882, Mr. Ridgway has described a considerable number of new species and races of birds and several new genera, chiefly from North and Middle America. They also contain notes on a few other hitherto little known species ...—J. A. A., *Bull. Nutt. Ornith. Club*, Vol. VIII., pp. 168, 169, July, 1883.

SAUNDERS, HOWARD.—On some Laridæ from the coasts of Peru and Chili, collected by Capt. Albert H. Markham, R.N., with Remarks on the Geographical Distribution of the Group in the Pacific. By Howard Saunders, F.L.S., F.Z.S. *Proc. Zool. Soc. of London*, June 6, 1882, pp. 520-530; with colored plate of *Xema furcatum* adult and young.

...Fifteen species are represented: among these is a specimen (the third one known) of *Xema furcatum*, now rediscovered after an interval of forty years' fruitless search Mr. Saunders is one of the few scientific writers who possess the happy faculty of making a technical treatise interesting to the average reader. The present paper.. .has a direct value to the student of North American ornithology, for much of its subject-matter... relates to species which are included in the North American Fauna.– W. B., *Bull. Nutt. Ornith. Club*, Vol. VIII., p. 54, January, 1883.

SHUFELDT, R. W.—Contributions to the Anatomy of Birds. By R. W. Shufeldt, M.D. [etc.] Author's edition, extracted (in advance) from the Twelfth Annual Report of the late U. S. Geological and Geographical Survey of the Territories (Hayden's). Washington: Government Printing Office, October 14, 1882. 8vo., title and pp. 593 806, pll. i-xxiv., many woodcuts in text.

It includes chapters on the osteology of *Speotyto cunicularia hypogæa*, *Eremophila alpestris*, the North American *Tetraonidæ* and the *Cathartidæ*. These subjects have been already treated by Dr. Shufeldt in previous papers....; but its subject-matter has been largely, if not entirely re-written, and some unfortunate errors .. corrected....The paper on the *Cathartidæ* with its accompanying plates, is entirely new matter. W. B., *Bull. Nutt. Ornith. Club*, Vol. VIII., p. 56, January, 1883.

Under this title a meritorious and very promising ornithotomist has brought together the greater part of what he has thus far accomplished in the way of avian anatomy... It would scarcely be fair, however, to judge their reappearance by their original character, all of them having been carefully revised and to some extent rewritten....The text is a faithful and on the whole an accurate description of the objects under designation, and the fidelity with which the plates are executed is most commendable ... E. C., *Bull. Nutt. Ornith. Club*, Vol. VIII. pp. 166-168, July, 1883.

STEJNEGER, LEONHARD.—Description of two new Races of Myadestes obscurus Lafr. By Leonhard Stejneger. *Proc. U. S. Nat. Mus.*, Vol. IV., 1882, pp. 371-374. Author's separates issued April 6, 1882.

...*M. obscurus* var. *occidentalis*, from the highlands of Southern Mexico and Guatemala, and *M. obscurus* var. *insularis*, from the Tres Marias Islands.—J. A. A., *Bull. Nutt. Ornith. Club*, Vol. VIII., p. 170, July, 1883.

WHEATON, J. M.—Report on the Birds of Ohio. By J. M. Wheaton, M.D. *Report of the Geological Survey of Ohio*, Vol. IV., pt. i., pp. 188–628. Columbus, Ohio: Nevins & Myers, State Printers. 8vo. 1882.

....a treatise on the ornithology of the State so extensive and so systematic that the time its preparation has occupied seems justified if not absolutely required .. Dr. Wheaton's report must at once take place at the head of State Faunas, so far as ornithology is concerned .. Ohioans have here, in fact, a correct history and description of their 300 birds, systematically arranged and classified, with diagnoses of the genera and higher groups, a considerable synonomy of each species with special reference to State literature, and a local bibliography.... this volume of some 450 pages is no slight nor uncertain addition to our ornithological literature. . .—E. C., *Bull. Nutt. Ornith. Club*, Vol. VIII., pp. 110–112, April, 1883.

White, George R., and Scott, W. L.—Commentary on the Bird-Fauna of the Vicinity of Ottawa. By Geo. R. White and W. L. Scott. Report of Ornithological and Oölogical Branch, *Trans. Ottawa Field Naturalists' Club*, No. 3, pp. 26–34, and Appendix.

. . The list is briefly annotated, and contains 169 species .. we are astounded to see in the list *Harporhynchus cinereus! Parus rufescens! Vireo pusillus! Glaucidium passerinum* var. *californicum*! This of course puts the whole affair under a cloud as an incompetent and doubtless pretty nearly worthless performance. -E. C., *Bull. Nutt. Ornith. Club*, Vol. VIII., p. 55, January, 1883.

[...The authors ...had no opportunity to correct the proof-sheets ....Edd.] *Bull. Nutt. Ornith. Club*, Vol. VIII., pp. 115, 116, April, 1883.

1883.

Beckham, Charles Wickliffe.—A List of the Birds of Bardstown, Nelson County, Kentucky. By Charles Wickliffe Beckham. *Journ. Cincinnati Soc. Nat. Hist.*, Vol. VI., pp. 136–147, July, 1883.

....the first paper on the birds of Kentucky, as such, which has yet appeared, and relating mainly to the birds of the immediate vicinity of Bardstown. "....no species has been admitted on any but the best of evidence; out of the one hundred and sixty-seven enumerated, the writer is himself responsible for all but eight of them" ...The list is briefly annotated . .is well printed, and evidently carefully prepared ....—J. A. A., *Bull. Nutt. Ornith. Club*, Vol. VIII., pp. 227, 228, October, 1883.

Cooke, W. W.—Mississippi Valley Migration. By W. W. Cooke. *Ornithologist and Oölogist*, Vol. VIII., Nos. 4–7, April–July, 1883, pp. 25–27, 33, 34, 41, 42, 49–53.

....Mr. Cooke's scheme contemplates a large number of observing stations ...he appears to have correspondents at 44 stations ...his matter is pertinent and in most cases well arranged; while his summaries respecting the movements of particular species, as given in his later papers, show at a glance what are the results attained.—J. A. A., *Bull. Nutt. Ornith. Club*, Vol. VIII., pp. 230, 231, October, 1883.

Cooke, W. W.—Bird Migration in the Mississippi Valley. By W. W. Cooke. *Forest and Stream*, Vol. XIX., Nos. 15, 16, 20, pp. 283, 284, 306, 384, November 9 and 16, and December 14, 1883.

Cory, Charles B.—Beautiful and Curious Birds of the World. By Charles B. Cory. Published by the Author. Part IV. Elephant folio. Three Plates, with Text.

. . contains plates of *Pseudogryphus californianus* . . . ; *Camptolaemus labradorius* . . . ; *Astrapia nigra*, the Incomparable Bird of Paradise . . . – W. B., *Bull. Nutt. Ornith. Club*, Vol. VIII., pp. 55, 56, January, 1883.

Cory, Charles B.—Beautiful and Curious Birds of the World. By Charles B. Cory. Published by the Author. Part V. Elephant folio. Three Plates, with Text.

. . . . has illustrations of *Epimachus magnus*, the Magnificent Bird of Paradise, *Epimachus ellioti*, Elliott's Bird of Paradise and *Pluvianus ægyptius*, the Crocodile Bird of the Nile. . . .– W. B., *Bull. Nutt. Ornith. Club*, Vol. VIII., pp. 55, 56, January, 1883.

Cory, Charles B.—Beautiful and Curious Birds of the World. By Charles B. Cory. Published by the Author. Parts VI. and VII. Elephant folio.

. . . . completes the work, which consists of twenty plates, with accompanying text. . . The plates . . are superb illustrations of some of the most striking forms of bird-life. . . . W. B., *The Auk*, Vol. I., p. 81, January, 1884.

Coues, Elliott.—A Hearing of Birds' Ears. By Elliott Coues. *Science*, Vol. II., Nos, 34, 38, and 39, pp. 422-424, 552-554, 586-589, Sept. 28, Oct. 26, Nov. 2, 1883, figg. 9.

. . . . A clear and detailed account of the mechanism of the ear in birds, taking the human ear as the chief basis of comparison . . . . – J. A. A., *The Auk*, Vol. I., p. 182, April, 1884.

Coues, Elliott, and Prentiss, D. Webster.– Bulletin of the United States National Museum, No. 26. Avifauna Columbiana : being a list of Birds ascertained to inhabit the District of Columbia, with the times of arrival and departure of such as are non-residents, and brief notices of habits, etc. The Second Edition, revised to date and entirely rewritten. By Elliott Coues, M.D., Ph.D., Professor of Anatomy in the National Medical College, etc., and D. Webster Prentiss, A.M., M.D., Professor of Materia Medica and Therapeutics in the National Medical College, etc. Washington : Government Printing Office. 1883. 8vo., pp. 133, 100 woodcuts, frontispiece, and 4 folded maps.

The title of this interesting brochure, although explicit, fails to fully imply the scope of the work. 4 pages of which are devoted to the 'Literature of the Subject,' 17 to the Location and Topography of

District,' 5 to the 'General Character of the Avifauna.' 78 to the 'Annotated List of the Birds,' 8 to a 'Summary and Recapitulation,' and 3 to the 'Game Laws of the District'....The original 'List'... published in 1862, contained 226 species ...The additions made in the twenty-two years which have intervened number 23 ...The subject in general is treated not only with great fulness, but is very attractively set forth, and in general plan forms an excellent model of what a faunal list should be ...—J. A. A., *The Auk*, Vol. I., p. 386, October, 1884.

GADOW, HANS.—Catalogue of the Birds in the British Museum. Vol. VIII. Catalogue of the Passeriformes, or Perching Birds. Cichlomorphæ: containing the Families Paridæ and Laniidæ (Titmice and Shrikes), and Certhiomorphæ (Creepers and Nuthatches). By Hans Gadow, Ph.D. London: Printed by order of the Trustees. 1883. 8vo., pp. i-xiii., 1-386, pll. i ix., and woodcuts in the text.

....Dr. Gadow's volume opens with the Paridæ (including the Regulidæ *auct.*), of which 10 genera and 82 species are recognized ....The Laniidæ embrace five subfamilies....The family Certhiidæ includes the Nuthatches as well as the Tree-Creepers ...In general, Dr. Gadow inclines to the recognition of comprehensive groups, from families downward. His reduction in genera and species from the hitherto current status is very marked....In method of execution, the present volume is strictly in accord with its predecessors, and is neither less valuable nor less welcome.—J. A. A., *The Auk*, Vol. I., pp. 279-281, July, 1884.

GILL, THEODORE.—Record of Scientific Progress for 1881. Zoölogy. By Theodore Gill. Smithsonian Report, 1881 (1883), pp. 408-498. Birds, pp. 481-490.

... a partial bibliography of noteworthy papers and works, and a synopsis of about half-a-dozen memoirs....—J. A. A, *The Auk*, Vol. I., p. 84, January, 1884.

GOSS, N. S.—A Catalogue of the Birds of Kansas. By N. S. Goss. Published under the direction of the Executive Council. Topeka, Kansas: Kansas Publishing House, 1883. 8vo., pp. iv., 34.

... a carefully annotated list of the birds of the State, prepared at the request and under the direction of the State Executive Council ... very few species are given on other authority than his own observations ....the list includes 320 species and races, 161 of which are marked as known to breed. The annotations are brief but pertinent....the list ....attains in general a high grade of excellence. ..—J. A. A., *Bull. Nutt. Ornith. Club*, Vol. VIII., p. 227, October, 1883.

JEFFRIES, J. AMORY.—The Epidermal System of Birds. By J. Amory Jeffries. *Proc. Boston Soc. Nat. Hist.*, Vol. XXII., pp. 203-240, pll. iv-vi. Dec., 1883.

...reports the results of his studies of the epidermal appendages in birds, with reference to their structure, development, and homologies....—J. A. A., *The Auk*, Vol. I., pp. 182, 183, April, 1884.

KING, F. H.—Economic Relations of Wisconsin Birds. By F. H. King. *Wisconsin Geological Survey*, Vol. I., chap. xi., pp. 441–610, figg. 103–114. Royal 8vo.

....Prof King's field-work....was commenced in 1873, and is apparently only just concluded his attention during this long period being steadily and rigidly directed to discovering what and how much food Wisconsin birds eat ...The facts recorded .. were obtained from an examination of the contents of over 1,800 birds ...The Introduction closes with "a Temporary Classification of Wisconsin Birds on an economic basis" .. The body of the report is primarily of the nature of an ordinary "local list" for the State of Wisconsin, giving in systematic order 295 species....The report is well written, giving in many cases extended biographies ...The numerous woodcuts are chiefly taken from Baird, Brewer, and Ridgway. –E. C., *Bull. Nutt. Ornith. Club*, Vol. VIII., pp. 107–110, April, 1883.

LAWRENCE, GEORGE N.—Descriptions of New Species of Birds of the Genera Chrysotis, Formicivora, and Spermophila. By George N. Lawrence. *Ann. New York Acad. Sci.*, Vol. II., 1882, No. 12, pp. 381–383. Issued June, 1883.

The species here described are: 1. *Chrysotis canifrons*....2. *Formicivora griseigula*....and 3. *Spermophila parva* ...–J. A. A., *The Auk*, Vol. I., p. 387, October, 1884.

MORTON, THOMAS, and ADAMS, CHARLES FRANCIS, JR.—The New English Canaan of Thomas Morton. With Introductory Matter and Notes by Charles Francis Adams, Jr. Boston: Published by the Prince Society. 1883. Sm. 4to., pp. vi, 381.—Chap. IV. Of Birds and Fethered Fowles, pp. 189–199. With notes by William Brewster and the Editor.

....reprinting Thomas Morton's "New English Canaan" (published originally in 1637), with editorial notes ...The technical notes on the birds, by Mr. Brewster, form an excellent commentary on the species mentioned by Morton ...Morton's New English Canaan, as thus admirably edited, includes nearly everything of interest bearing upon the natural history of New England contained in these early accounts of New England ...The work is limited to 250 copies, and in typography and paper is a noteworthy specimen of book-making.—J. A. A., *The Auk*, Vol. I., p. 81, January, 1884.

NELSON, E. W.—Birds of Bering Sea and the Arctic Ocean. By E. W. Nelson. Cruise of the Revenue-steamer Corwin in Alaska and the N. W. Arctic Ocean in 1881. Notes and Memoranda: Medical and Anthropological; Botanical; Ornithological. Washington: Government Printing Office. 1883. 1 vol., 4to., pp. 55, 56, 56*a–f*, 57–118; with 4 colored plates.

...It is a pity that so valuable and interesting a treatise as this of Mr. Nelson's should not have been more carefully printed ...After some pages concisely descriptive of the region and its avifauna, the author proceeds to treat, in more or less detail, no fewer than 192 species of birds, North American with few exceptions....it is illustrated with four colored plates, executed by Mr. [Robert] Ridgway, representing *Motacilla ocularis*, *Lanius cristatus*, *Eurynorhynchus pygmaeus* and *Cireronia pusilla* ....—E. C., *The Auk*, Vol. I., pp. 76-81, January, 1884.

Ridgway, Robert.—A Review of the American Crossbills (Loxia) of the L. curvirostra type. By Robert Ridgway. *Proc. Biol. Soc. of Washington*, Vol. II., 1883, pp. 84-107.

....He recognizes three races of American Red Crossbills, one of which (*L. cuvirostra bendirei*) is described as new ...In North America the Red Crossbills decrease in size from the north southward ..There are also remarks on other races of Red Crossbills, particularly the *L. curvirostra* and *L. pityopsittacus* of Europe.—J. A. A., *The Auk*, Vol. II., pp. 206, 207, April, 1885.

Ridgway, Robert—Description of Several new Races of American Birds. By Robert Ridgway. *Proc. U. S. Nat. Mus.*, Vol. V., 1883, pp. 9-15. Author's separates issued June 5, 1882.

Ridgway, Robert—On the genera Harporhynchus, Cabanis, and Methriopterus, Reichenbach, with a description of a new genus of Miminæ. By Robert Ridgway. *Proc. U. S. Nat. Mus.*, Vol. V., 1883, pp. 43-46. Author's separates issued June 5, 1882.

Ridgway, Robert—Critical Remarks on the Tree-creepers (Certhia) of Europe and North America. By Robert Ridgway. *Proc. U. S. Nat. Mus.*, Vol. V., 1883, pp. 111-116. Author's separates issued July 8, 1882.

....he proceeds to characterize seven races as susceptible of definition, three of which are for the first time named....—J. A. A., *Bull. Nutt. Ornith. Club*, Vol. VIII., p. 113, April, 1883.

Ridgway, Robert—Description of some new North American Birds. By Robert Ridgway. *Proc. U. S. Nat. Mus.*, Vol. V., 1883, pp. 343-346. Author's separates issued Sept. 5, 1882.

Ridgway, Robert—Catalogue of a Collection of Birds made in the Interior of Costa Rica, by Mr. C. C. Nutting. By Robert Ridgway. *Proc. U. S. Nat. Mus.*, Vol. V., 1883, pp. 493-502. Author's separates issued Feb. 28, 1883.

The collection reported upon was made partly at Volcan de Irazú and partly at San José....There are brief field-notes by the collector, and technical notes on a few species by Mr. Ridgway.—J. A. A., *The Auk*, Vol. I., p. 84, January, 1884.

Ridgway, Robert.—Description of a New Warbler, from the Island of Santa Lucia, West Indies. By Robert Ridgway. *Proc. U. S. Nat. Mus.*, Vol. V., 1883, pp. 525, 526. Author's separates issued March 21, 1883.

Mr. Ridgway separates as a new subspecies the Warbler from Santa

Lucia, W. I., hitherto known as *Dendroica adelaidæ* under the name of *Dendroica adelaidæ delicata* ... J. A. A., *The Auk*, Vol. I., p. 83, January, 1884.

RIDGWAY, ROBERT.—Description of a supposed New Plover, from Chili. By Robert Ridgway. *Proc. U. S. Nat. Mus.*, Vol. V., 1883, pp. 526, 527. Author's separates issued March 21, 1883.

(*Ægialites albidipectus*, sp. nov.) based on a single example from Chili. J. A. A., *The Auk*, Vol. I., p. 83, January, 1884.

RIDGWAY, ROBERT.—On the Genus Tantalus, Linn., and its allies. By Robert Ridgway. *Proc. U. S. Nat. Mus.*, Vol. V., 1883, pp. 550, 551. Author's separates issued March 21, 1883.

The genus *Tantalus* Linn., is restricted to *T. loculator*, ...—J. A. A., *The Auk*, Vol. I., p. 83, January, 1884.

RIDGWAY, ROBERT.—Description of a New Petrel from Alaska. By Robert Ridgway. *Proc. U. S. Nat. Mus.*, Vol. V., 1883, pp. 656-658. Author's separates issued June 26, 1883.

....(*Æstrelata fisheri*, sp. nov.) from Alaska, a species most nearly allied to *Œ. defillipiana* Mr. Ridgway is inclined to refer also the Petrel taken in Livingston County, N. Y., identified by Mr. Brewster ... as *Œ. gularis*, to *Œ. fisheri*. J. A. A., *The Auk*, Vol. I., p. 83, January, 1884.

RIDGWAY, ROBERT.—Notes upon some Rare Species of Neotropical Birds. By Robert Ridgway, Curator Department of Birds, United States National Museum. *The Ibis*, October, 1883, pp. 399–401.

The species considered are *Harporhynchus ocellatus* Scl., *Pyranga erythrocephalus* (Sw.), *Zonotrichia quinquestriata* Scl. & Salv., *Contopus ochraceus* Scl. & Salv., and *Panyptila cayennensis* (Gm.), about which there are brief remarks respecting their affinities .... J. A. A., *The Auk*, Vol. I., pp. 386, 387, October, 1884.

SEEBOHM, HENRY.—A History of British Birds, with colored Illustrations of their Eggs. By Henry Seebohm. London: Published for the author by R. H. Porter, 6 Tenterden Street, W., and Dulau & Co., Soho Square, W. Royal 8vo., Vol. I., 1883, pp. xxiv., 613, pll. 20; Vol. II., (Part 1, 1883, Part 2, 1884) pp. xxxiv., 600, pll. 22.

....The typographical execution of the work is excellent, and the plates are entitled to high praise....In respect to nomenclature and classification Mr. Seebohm is conservative to a degree approaching eccentricity, but in respect to the general subject his views are liberal, philosophic, and progressive....As regards classification Mr. Seebohm seems inclined to ignore all recent progress, ...In respect to the 'vexed question of nomenclature' he has throughout his work "set the Rules of the British Association at defiance...." His panacea for the evil is... the adoption of an '*auctorum plurimorum*' rule;... For subspecies he adopts what may be termed a Seebohmian system of trinomials first instituted by him in his British Museum Catalogue of the Turdidæ ...As Mr. Seebohm says: "The real history of a bird is its *life*-history. The deepest interest attaches to every thing that reveals the little *mind*, however feebly it may be developed, which lies behind the feathers. The habits of the bird during the breeding season, at the two periods of migration, and in winter; its mode of flight and of progression on the ground, in the trees, or on the water; its song and its various call- and alarm-notes; its food and its means of procuring it at different seasons of the year; its migrations, the dates of arrival and departure, routes it chooses, and the winter quarters it selects; and above all, every particular respecting its breeding, when it begins to build its nest, the materials it uses for the purpose, the number of eggs it lays, the variations in their color, size and shape, all these particulars are the real history of a bird; and in the account of each species of British birds I endeavor to give as many of them as possible."....Mr. Seebohm's work abounds in passages which invite comment ...—J. A. A., *The Auk*, Vol. II, pp. 88-91, January, 1885.

SHARPE, R. BOWDLER.—Catalogue of the Birds in the British Museum. Vol. VII. Catalogue of the Passeriformes, or Perching Birds. Cichlomorphæ: Part IV., containing the concluding portion of the Family Timeliidæ (Babbling Thrushes). By R. Bowdler Sharpe. London: Printed by order of the Trustees. 1883. 8vo., pp. i-xvi., 1-698, pll. i-xv., and numerous woodcuts in the text.

The family *Timeliidæ*, an account of which was commenced in the preceding volume [Vol. VI.], is here [Vol. VII.] completed, with the enumeration and description of 687 species....while many ornithologists may not agree with the author in his allocation of certain forms, none, we fancy, can feel otherwise than deeply grateful to him for the very useful monograph he has placed at their disposal.—J. A. A., *The Auk*, Vol. I., pp. 278, 279, July, 1884.

SMITH, EVERETT.—The Birds of Maine. With annotations of their comparative abundance, dates of migration, breeding habits, etc. By Everett Smith. *Forest and Stream*, Vol. XIX., Nos. 22-26; Vol. XX., Nos. 1-7 and 10-13.

....Passing to water birds it is gratifying to find a better quality of work. Mr. Smith is evidently at home here, and proofs of the general accuracy of his information and judgement are numerous and unmistakable....It is too good a paper to be wholly condemned, too faulty a one to be generously praised .. Its author... has become almost an ornithologist....—W. B., *Bull. Nutt. Ornith. Club*, Vol. VIII., pp. 164-166, July, 1883.

STEARNS, W. A.—Notes on the Natural History of Labrador. By W. A. Stearns. *Proc. U. S. Nat. Mus.*, Vol. VI., 1883, pp. 111-137. Author's separates issued July 27 to Sept. 20, 1883.

These "Notes" relate only in part to birds, which occupy pp. 116-123. The list of birds numbers 111 species, and is briefly annotated....—J. A. A., *The Auk*, Vol. I, p. 284 July, 1884.

STEARNS, WINFRID A., and COUES, ELLIOTT.—New England Bird Life: being a Manual of New England Ornithology. Revised and edited from the manuscript of Winfrid A. Stearns, Member of the Nuttall Ornithological Club, etc. By Elliott Coues, Member of the Academy, etc. Part II. Non-oscine Passeres, Birds of Prey, Game and Water Birds. Boston: Lee & Shepard, Publishers. New York: Charles T. Dillingham. 1883. 8vo., pp. 409; 88 woodcuts.

....Dr. Coues has gone bravely on with the task of "editing" Mr. Stearns's manuscript, and the finished work, now complete in two volumes, is the gratifying result. Much that we said... of Part I. will apply equally to Part II.... But among the Water Birds there are rather frequent evidences of hasty, and often positively incorrect conclusions ....New England Bird Life...., is, on the whole a wisely-conceived and admirably-executed book—by far the best, in fact, which has been so far published on New England birds ... W. B., *Bul. Natt. Ornith. Club*, Vol. VIII., p. 161-164. July, 1883.

STEJNEGER, LEONHARD.—Synopsis of the West Indian Myiadestes. By Leonhard Stejneger. *Proc. U. S. Nat. Mus.*, Vol. V., 1883, pp. 15-27, pl. ii. Author's separates issued June 5, 1882.

Eight species are recognized, two of which *M. sanctæ-luciæ*, *M. dominicanus*) are described as new.—J. A. A., *Bull. Natt. Ornith. Club*, Vol. VIII., p. 170, July, 1883.

STEJNEGER, LEONHARD.—On some generic and specific appellations of North American Birds. By Leonhard Stejneger. *Proc. U. S. Nat. Mus.*, Vol. V., 1883, pp. 28-43. Author's separates issued June 5, 1882.

Proposing to use "the oldest available name in every case" the author shows that many of our current names must give way if the "inflexible law of priority" is to be observed. For ourselves we believe that the surest way out of the nomenclatural difficulties that beset us is to be found in some such simple rule as this .... Still such a paper as this makes us wish ... that some counteractive "statute of limitation" could come into operation .... Stejneger's points seem to be well taken in the main; and .... we presume the restrictions and substitutions he proposes are available if not indeed necessary under the priority statute ... E. C., *Bull. Natt. Ornith. Club*, Vol. VII., pp. 178, 179, July, 1882.

STEJNEGER, LEONHARD.—Outlines of a Monograph of the Cygninæ. By Leonhard Stejneger. *Proc. U. S. Nat. Mus.*, Vol. V., 1883, pp. 174-221, figg. 16. Author's separates issued July 25, 1882.

The external and osteological characters are given in detail, with diagnoses of the genera and species ... the author recognizes four genera of Swans, namely *Sthenelus* (gen. nov.), *Cygnus*, *Olor*, and *Chenopsis*. The two North American species are assigned to *Olor*. J. A. A., *Bull. Natt. Ornith. Club*, Vol. VIII., p. 231, October, 1883.

Stejneger, Leonhard.—Remarks on the Systematic Arrangement of the American Turdidæ. By Leonhard Stejneger. *Proc. U. S. Nat. Mus.*, Vol. V., 1883, pp. 449–483, with numerous cuts. Author's separates issued February 13, 1883.

....Dr. Stejneger's synopsis of the family extends only to the genera and higher groups as represented in America. The generic synonomy is fully given, and the generic diagnoses are supplemented by general remarks and figures illustrative of the principal generic characters.—J. A. A., *The Auk*, Vol. I., pp. 181, 182, April, 1884.

Townsend, Charles H.—Notes on the Birds of Westmoreland County, Penna. By Charles H. Townsend. *Proc. Acad. Nat. Sci. Philadelphia*, 1883, pp. 59–68.

"The species enumerated represent perhaps not more than two-thirds of the actual birds of Westmoreland County"....The list, numbering 136 species, is rather too sparingly annotated ...but we are led to hope that this may be the forerunner of a fuller report.—J. A. A., *The Auk*, Vol. I, p. 184, April, 1884.

Tuelon, James A.—List of Birds observed near Bradford, Pa., by James A. Tuelon. *Quarterly Jour. Boston Zoöl. Soc.*, Vol. IV., January, 1883, pp. 8–11.

As the whole number is only 77, without exception very common and well-known species, and as the annotations are of no special consequence, the reason why the list is printed is not evident.—E. C., *Bull. Nutt. Ornith. Club*, Vol. VIII., p. 171, July, 1883.

Turner, Lucien M.—On Lagopus mutus, Leach, and its Allies. By Lucien M. Turner. *Proc. U. S. Nat. Mus.*, Vol. V., 1883, pp. 225–233. Author's separates issued July 29, 1882.

The author believes....that the European birds *mutus* and *alpinus* constitute "but a single species having the name *Lagopus mutus* Leach, while the American bird ...to be called *Lagopus mutus rupestris* (Gm.) Ridg. Four races are recognized ...Detailed descriptions and measurements are given of a considerable number of specimens of each race. —J. A. A., *Bull. Nutt. Ornith. Club*, Vol. VIII., p. 232, October, 1883.

Willard, S. W.—Migration and Distribution of North American Birds in Brown and Outagamie Counties. By S. W. Willard. De Pere, Wis., 1883, 8vo., pp. 20.

....The paper gives evidence of careful observation, and is a valuable contribution to our knowledge of the manner of occurrence and movements of the birds of the area in question.—J. A. A., *The Auk*, Vol. II., pp. 289, 290, July, 1885.

Note.—Publication of Part II. of this paper is deferred to a succeeding number of these 'Abstracts.'

www.ingramcontent.com/pod-product-compliance
Lightning Source LLC
Chambersburg PA
CBHW031759090426
42739CB00008B/1081

9783337241544